Christian Hintze

Influence of processing induced morphology on mechanical properties of short aramid fibre filled elastomer composites

TUD*press*

Christian Hintze

Influence of processing induced morphology on mechanical properties of short aramid fibre filled elastomer composites

TUDpress
2013

Die vorliegende Arbeit wurde am 22. Oktober 2012 an der Fakultät Maschinenwesen der Technischen Universität Dresden als Dissertation eingereicht und am 19. März 2013 erfolgreich verteidigt.

Gutachter:
Prof. Dr. rer. nat. habil. Gert Heinrich
Prof. Jacques W. M. Noordermeer, Ph.D., M.Sc., B.Sc.

The studies described in this thesis are part of the Research Programme of the Dutch Polymer Institute, P.O. Box 902, 5600 AX Eindhoven, The Netherlands, project nr. 664.

Bibliografische Information der Deutschen Nationalbibliothek
Die Deutsche Nationalbibliothek verzeichnet diese Publikation in der Deutschen Nationalbibliografie; detaillierte bibliografische Daten sind im Internet über http://dnb.d-nb.de abrufbar.

Bibliographic information published by the Deutsche Nationalbibliothek
The Deutsche Nationalbibliothek lists this publication in the Deutsche Nationalbibliografie; detailed bibliographic data are available in the Internet at
http://dnb.d-nb.de.

ISBN 978-3-944331-13-3

Verlag der Wissenschaften GmbH
Bergstr. 70 | D-01069 Dresden
Tel.: 0351/47 96 97 20 | Fax: 0351/47 96 08 19
http://www.tudpress.de

Gesetzt vom Autor.
Printed in Germany.

Influence of processing induced morphology on mechanical properties of short aramid fibre filled elastomer composites

An der Fakultät Maschinenwesen
der Technischen Universität Dresden
zur Erlangung des akademischen Grades

Doktoringenieur
(Dr.-Ing.)

eingereichte

Dissertation

vorgelegt von

Dipl.-Ing. Christian Hintze

geboren am 29.09.1982
in Karl-Marx-Stadt

Gutachter:
Prof. Dr. rer. nat. habil. Gert Heinrich
Prof. Jacques W. M. Noordermeer, Ph.D., M.Sc., B.Sc.

Tag der Einreichung: 22.10.2012

I wonder about the apparent victory of our selfish, evolutionary short-shighted reasoning that maximes personal goods at present over the intellectual or moral rationality that would have been able to avoid the crisis.

Dag Olav Hessen

Danksagung

Die vorliegende Arbeit entstand während meiner Tätigkeit als wissenschaftlicher Mitarbeiter am Leibniz-Institut für Polymerforschung Dresden e.V. (IPF). Wesentliche Teile der Arbeit entstanden im Rahmen des Dutch Polymer Institute (DPI) Projektes SUSTFIBER "Sustainable elastomers and thermoplastics by short-cut fiber reinforcement".

Mein besonderer Dank gilt Herrn Prof. Gert Heinrich für die Betreuung und Unterstützung der Arbeit. Prof. Jacques W. M. Noordermeer sei für die Übernahme der Gutachtertätigkeit und seine Führung als Projektleiter gedankt.

Weiterhin danke ich insbesondere Jun.-Prof. Sven Wießner für seine Unterstützung und stete Diskussionsbereitschaft, Regine Boldt, Prof. Udo Wagenknecht sowie allen Kollegen der Abteilung Verarbeitungsprozesse und der Schweizer Strasse für ihre individuelle Unterstützung und das gute Arbeitsklima. Morteza Sadat Shirazi und Natalia Rodriguez, meinen Mitdoktoranden im Projekt, möchte ich für die gute Zusammenarbeit danken. Ganz herzlich möchte ich mich bei allen Studenten, namentlich Michael Spaethe, Katja Eisleben, Christopher Drosdzol und Romy Schumann bedanken, die durch ihre Arbeiten zum Gelingen der Arbeit beitrugen.

Dank gilt auch der Elastomergruppe am IPF, besonders Dr. K. W. Stöckelhuber, Dr. Amit Das und René Jurk für ihre Hilfestellung bei verschiedenen praktischen und theoretischen Fragestellungen, sowie Dr. Thomas Horst für die Überzeugungsarbeit diese Dissertation in Tex zu schreiben. Für die Hilfe mit ImageJ bedanke ich mich bei Dr. Sven Pegel und für die Hilfe am Konfokalmikroskop bei Martin Kaufmann. Dank auch an Dr. Hannes Michael und Dr. Radek Stocek von der TU Chemnitz für die unkomplizierte Unterstützung in Form von Kautschukmustern und Messungen am Tear Fatigue Analyser. Kiitos an Prof. Jyrki Vuorinen und Katja Nevalainen von der TU Tampere für ihre Gastfreundschaft und die Möglichkeit meine Diplomarbeit in Finnland anzufertigen.

Weiter sei Teijin Aramid B.V. für das Bereitstellen der Kurzschnittaramidfasern, und besonders Dr. Pieter J. de Lange für die Unterstützung und Diskussionsbereitschaft gedankt. Ebenso möchte mich für die Unterstützung von DSM Elastomers N.V. mit EPDM, Masterbatchsystemen und den Vulkansisationschemikalien bedanken. Für die NanoCT Messung Dank an die Phoenix x-ray GmbH. Dank auch an das Dutch Polymer Institute für die finanzielle Förderung dieser Arbeit.

Abschließend bedanke ich mich bei meiner Familie und meinen Freunden für den Rückhalt und ihre Geduld in dieser arbeitsreichen Zeit.

Dresden, im Jahre 2012 Christian Hintze

Zusammenfassung

In dieser Arbeit wurde der Einfluss verfahrenstechnischer und stofflicher Parameter auf Morphologie und Anwendungseigenschaften aramidkurzfaserverstärkter Elastomercompounds untersucht. Speziell die verfahrensabhängige Dispersion, Orientierung, Längenverteilung und Faser-Matrix-Haftung zweier unterschiedlichen Typen von Aramidkurzfasern mit verschiedenen Schlichten in Ethylen-Propylen-Dien-Kautschuk (EPDM) und Naturkautschuk (NR) und deren Einfluss auf die Anwendungseigenschaften der resultierenden Elastomerkomposite wurde bestimmt.

Dazu wurden zwei Fasertypen mit unterschiedlichen Schlichten und Ausgangslängen im ersten Schritt auf einem Laborinnenmischer oder einem Walzwerk eingemischt. Bevor die Compounds in einem Presswerkzeug vulkanisiert wurden, arbeitete man auf dem Walzwerk in einem zweiten Schritt die Vulkanisationssysteme ein und orientierte die Fasern gleichzeitig gezielt. Da die meisten lichtoptischen Charakterisierungsmethoden durch den in industrietypischen Rezepturen eingesetzten Ruß behindert werden, wurde für die Morphologiecharakterisierung zu jedem rußgefüllten Compound ein korrespondierendes transparentes Modellsystem hergestellt.

Zur Charakterisierung der Komposite wurden Licht- und Elektronenmikroskopie sowie quasistatische und dynamische Werkstoffprüfungen genutzt. Da es sich bei Aramidfasern um fibrillierende, flexible Fasern in hochelastischen Matrices handelte, musste die Probenpräparation und die Charakterisierungsmethoden zur Faserlängenbestimmung (Fluoreszenzmikroskopie), und zur Bestimmung der Orientierung (Konfokalmikroskopie) erst erarbeitet und getestet werden. Die Dispersion und Längenverkürzung der Fasern in den Modellsystemen zeigte eine starke Abhängigkeit von der Deformation im Mischprozess und wurde außerdem vom verwendeten Fasertyp, der Schlichte sowie der Matrix beeinflusst. Aus den Ergebnissen der Zugversuche konnte in Faserrichtung eine Streckgrenze im Bereich kleiner Dehnungen nachgewiesen werden. Mit zunehmender Faser-Matrix-Grenzfläche stiegen die Spannungswerte an. Die Grenzfläche war dabei vom Fasergehalt und von der verbleibenden Faserlänge abhängig. Waren die Fasern senkrecht zu Zugrichtung orientiert, blieb die Verstärkungswirkung gering. Durch die Ermittlung der Faserlänge und der Orientierung nach der Verarbeitung konnten die Elastizitätsmoduln der Komposite theoretisch vorausgesagt werden. Die Dynamisch Mechanische Analyse zeigte korrespondierende Ergebnisse. Mit zunehmender Faser-Matrix-Grenzfläche nahmen die Maxima des Verlustfaktors in Faserrichtung ab. Ein direkter Beweis für eine erhöhte Faser-Matrix-Haftung durch den Einsatz von Schlichten konnte nicht erbracht werden, da die Grenzfläche ab einer bestimmten Dehnung versagte. Die gefundenen Morphologie-Eigenschafts-Beziehungen konnten im Weiteren auf die rußgefüllten Komposite übertragen werden und führten zu einem besseren Verständnis von dessen Einflussfaktoren.

Abstract

The aim of the present work was to investigate and correlate the influences of processing and material parameters on the morphology and resulting properties of short aramid fibre elastomer composites. Especially the dispersion, length distribution and fibre-matrix interaction of two types of different dipped short aramid fibres in ethylene-propylene-diene rubber (EPDM) and natural rubber (NR) and their influence on the resulting properties was under investigation.

Two fibre types with different surface treatments and initial fibre lengths were incorporated in a first step on laboratory scale two roll mill or internal mixer. Before the compounds were vulcanised by compression moulding, mixing in of curatives and orienting of the fibres were carried out in a second step on two roll mill. Because most optical microscopy techniques were hampered by the presence of carbon black in typical industrial formulations, corresponding transparent model compounds were prepared to determine the fibre morphology.

Optical- and electron microscopy, tensile testing and dynamic mechanical testing were applied to explore the short aramid fibre reinforced elastomer composites. Since aramid fibres are very flexible and tend to fibrillate during mixing while the matrices are highly elastic, the characterisation methods used to determine the fibre length distribution (fluorescence microscopy) and orientation distribution (Confocal laser scanning microscopy) had to be developed and tested.

The fibre dispersion and length reduction in the model systems was mainly dependent on the specific mixing energy and in addition on the used fibre type, -dip and matrix. The results of tensile testing in fibre direction showed a yield point at small strains. With growing fibre-elastomer interface the stress values were increasing. The interface was dependent on the fibre content and residual length.

If the fibres were oriented perpendicular to the direction of tensile testing the reinforcement level was rather low. Due to the determination of the fibre length and orientation after processing the corresponding Young's modulus could be calculated theoretically. For the dynamic mechanical testing similar results were found. With rising fibre-elastomer interface the maxima of the measured loss factor declined when measured in fibre direction.

A direct proof for an enhanced fibre matrix adhesion through the presence of fibre surface treatment could not be confirmed since the interface failed above a certain strain level in form of fibre pull-out. The determined morphology-property relationships for the transparent compounds were successfully transferred to the carbon black filled compounds and were therefore leading to better understanding of their influencing factors.

Contents

Chapter 1

Introduction

1.1 Motivation

The limits of growth and their consequences on population, resources, food, industrial output and pollution on earth were modelled and discussed by Meadows et al. [1]. Despite the fact that the authors stated already in 1972 that with exponential growth of population and economy by simultaneously finite global resources the growth will have physical limits and the existing system will collapse if not changed, this message is quite up to date. The following quotation from the book illustrates some of the important consequences:

> A transition to sustainability will require, however, an active decision to reduce the human ecological footprint. This, in turn, will require personal decisions to reduce family size, lower goals for industrial growth, and raise efficiency in use of the earth's resources

According to Wackernagel et al. [2] the overshot is already reached since the 1980s. During the last years some indicators were observable that the limits of the earth are already reached or overshot. The rapid increase of world population to seven billion as well as the advancing industrialisation of the emerging countries leads to dramatic growth of needed resources. At the same time the critical resources in the areas energy, agriculture and base metals are limited, which was leading to price peaks in 2008 and 2011. This increase of commodity prices also caused partially the financial crises starting in 2008 and with delay the Euro crisis. Beside these economic problems the already high demand of resources, especially in energy area, was leading to the acceptance of higher risks. The results were a rise of greenhouse gas emissions levels and irreversible pollution with the outstanding examples "Deep water horizon oil spill" in 2010 and "Fukushima" in 2011.

Also the rubber and tyre industry is nowadays facing the general increase of raw materials [3] as the customers are confronted with rising fuel prices [4]. Therefore in November 2012 the European Union (EU) introduced a tyre label [5] as shown in figure 1.1 showing fuel consumption, wet grip and noise classification of the corresponding tyres. According to industry a tyre in use, corresponding to its rolling resistance, stands for roughly 90% of its CO_2 footprint [6]. Therefore it should be the target to reduce the rolling resistance while maintaining good wet grip and low wear at the same time. This is not an easy goal since these properties cannot be tuned without affecting each other's performance in the so called "magic triangle". In tyres short fibres can lower the fuel consumption by reducing the rolling resistance and improve the abrasion resistance [7].

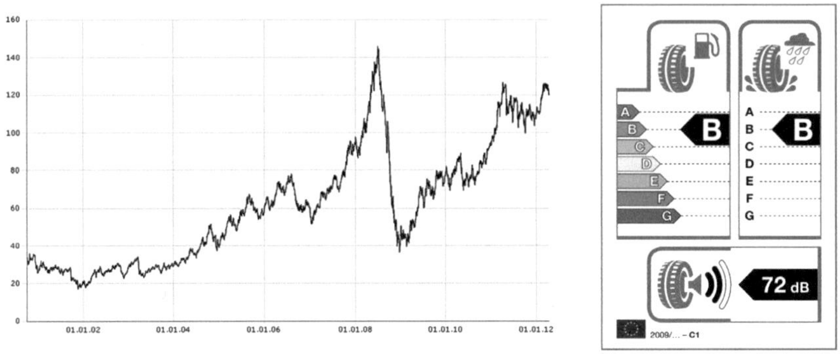

Figure 1.1: Brent crude oil price in US dollar 2000–2012 (left [4]), new EU tyre label introduced in November 2012 (right [5]),

Generally short fibres mixed into elastomers can contribute to the improvement of hardness, heat build-up and wear properties and therefore increase the durability of the corresponding applications. Especially the improvement of crack deflection and the achievable anisotropic properties are an advantage of these high aspect ratio fillers over particulate fillers like carbon black or silica. Short fibre reinforcement in elastomers can provide a substantial contribution to build more efficient and durable tyres, transmission- and conveyor belts, seals and hoses.

1.2 Objectives and structure of the thesis

The main objective of this thesis is to study the influence of processing induced morphology of short aramid fibre filled elastomer composites. Because the short fibres can be mixed in the elastomer compounds by standard rubber processing steps together with other additives, the throughput is higher compared to continuous fibre reinforced elastomers. However, this advantage can just be achieved, when the following influencing factors are controlled in the right manner:

- Dispersion of bundles down to filament level
- Keeping a high aspect ratio during compounding and processing
- Controlled orientation in desired direction for application
- Sufficient fibre-elastomer interaction

Since several fibre types are not able to withstand the high shear forces occurring during mixing of elastomer compounds, two aramid fibres with different breakage behaviour were chosen to study the influence of fibre length on the composite properties. The use of short aramid fibres is generally hampered by the chemical inertness of the aramid surface, leading to a lack of interaction between fibre and elastomer, resulting in insufficient fibre-matrix adhesion. Therefore several treatments were applied to the fibres. The influence of these fibre dips on the dispersion and residual fibre length was characterised and correlated with the final properties.

In most of the published works the use of short fibres was investigated in typical elastomer compounds containing carbon black and various other components. However, this filler hampers the morphology detection and makes it sometimes impossible to differentiate the contributions of fibre dispersion, -length and -orientation and the fibre-elastomer interface. Therefore peroxide cured model compounds without carbon black were chosen to achieve transparency and correlate the fibre morphology directly with the final mechanical, dynamic and wear properties of the composites. Due to the transparency of the compounds it was also possible to determine the main fibre orientation and predict the measured values by theories of short fibre reinforcement.

This thesis is structured in the following way. Introduction of topic, motivation, objectives and structure of the thesis is presented in the first chapter.

In the second chapter the state of the art is reported for short fibre filled elastomers compounds. The theory of short fibre reinforcement is described and the typically used elastomers and short fibres are introduced in the following. The focus is hereby directed especially on short aramid fibres and their structure, properties and failure modes. Later the influence of processing parameters, possibilities to enhance the fibre-elastomer interaction, the typically applied characterisation methods, resulting composite properties, final applications and claimed patents are presented.

The used compound components, processing steps and devices, characterisation methods and standards are described in the third chapter.

In chapter four the gained results are discussed, correlated and modelled by theories for short fibre reinforcement. The first subchapter deals with the basic properties of the used fibres and elastomers and is followed by a subchapter where the results of the compounding process and dispersion are presented. The influence of the different material and processing parameters on the fibre aspect ratio is characterised in the next subchapter. Next subchapter deals with the curing behaviour of the compounds. Since the orientation has a major influence on the composite properties the degree of orientation was investigated depending on several fibre and processing parameters. In the following the mechanical and dynamic behaviour was reported and in case of the model systems correlated and modelled with the found fibre morphology. In the last two subchapters the fatigue behaviour of the carbon black filled compounds was characterised in terms of heat build-up and tear fatigue analysis.

Finally, the summary, conclusions and an outlook are presented in chapter five.

Chapter 2

State of the Art

The present chapter provides an overview of literature published on short fibre reinforced elastomer composites. A brief review of used matrices and fibres in rubber applications is given. The emphasis is on elastomer composites reinforced with short aramid fibres. The processing and compounding of short fibre reinforced rubber compounds, used characterisation methods, properties and applications are described.

Reinforcement of rubber products by fibres is an established standard in industry. In the past continuous cords have been used for this reason. Their orientation can be easily fixed to optimise performance in the direction of the applied load. The throughput of these continuous cord reinforced rubber products is lower compared to short fibre reinforced elastomers [8]. The possibility to incorporate short fibres directly into the rubber compound during mixing has been the reason why short fibres gained more and more importance over the years. Improved tear, wear and penetration resistance and controlled hysteresis as desired properties have also contributed to this development. On the other hand some issues have to be taken into account: preservation of a high aspect ratio of the fibre during processing, anisotropic properties caused by fibre orientation, creation of sufficient interfacial adhesion between fibre and matrix, optimal dispersion of the fibres in the matrix and optimised rubber formulation as the base of the compound. Short fibre reinforced rubbers are used in tyres, transmission and conveyor belts, hoses and seals [7,9]. Some literature about short fibre reinforcement of rubbers has been published. De and White [9] reviewed short fibre reinforced polymers in general. Papathanasiou [10] reviewed flow induced alignment in polymer composite materials. Reinforcement of rubber by short fibres and continuous fibres was reviewed by Rajeev [11]. Goettler and Cole [12] reviewed short fibre reinforcement of rubbers in general, while Abrate [13] reviewed the mechanics of short fibre reinforced rubbers.

2.1 Theory of short fibre reinforcement

In this section the theories for reinforcement of mechanical properties of short fibre reinforced composites are presented. The stress distribution in a reinforcing fibre is examined as well as the factors that influence the mechanical properties of short fibre reinforced elastomers.

2.1.1 Stress transfer

When short fibres are used in elastomer composites, the applied load is transferred from the elastomer to the fibre. To have effective composites the load has to be transferred by shear forces at the fibre-elastomer interface from the matrix into the fibre. Assuming a perfect bond between the fibre and the elastomer, the difference in strains along the fibre creates a shear stress distribution across the fibre-elastomer interface. As the strain in the composite increases, the strain in the fibre increases from the fibre ends. The stress increases from zero at the fibre end and reaches a plateau value along the middle of the fibre [14]. A suitable stress transfer model was proposed by Rosen [15] with his shear lag theory. He assumed that the effect of adjacent fibres on the stress distribution and the effect of the fibre end geometry can be ignored. In this condition, maximum reinforcement would be achieved when the fibres are long enough so that complete stress transfer occurs. So it is possible to calculate a minimum critical length, which is required for obtaining complete stress transfer, based on the equation proposed by Rosen:

$$\frac{l}{d} = \left[\frac{1}{2} \times \frac{E_f}{G_m} \times \frac{1-\phi_f^{1/2}}{\phi_f^{1/2}}\right]^{1/2} \tag{2.1}$$

where l/d is the fibre aspect ratio, E_f is the elastic modulus of the fibre, G_m is shear modulus of the matrix and ϕ_f is the fibre volume fraction. The critical fibre length, for which complete stress transfer can be achieved is called effective aspect ratio.

To keep high fibre aspect ratios near the critical length is quite hard, since there is significant fibre breakage taking place during the steps of rubber processing. Complete stress transfer in short fibre reinforced elastomers is difficult, since the elastic properties regarding stiffness and strain at break are so different, that these composite will either fail in the soft elastomer matrix or in the interface [16].

2.1.2 Fibre-fibre interaction

It is necessary to characterise the fibre-fibre interactions, because of its impact on melt flow and thus the resulting morphology in the vulcanisate. According to Papathanasiou [10] fibre suspensions are classified in three concentration regimes:

$$\phi_f < \frac{1}{{r_f}^2} \text{ dilute} \tag{2.2}$$

$$\frac{1}{{r_f}^2} < \phi_f < \frac{1}{r_f} \text{ semi-concentrated} \tag{2.3}$$

$$\frac{1}{r_f} < \phi_f \text{ concentrated} \tag{2.4}$$

where ϕ_f is the fibre volume fraction and r_f is the fibre aspect ratio defined as l/d.

In the dilute region, the spacing between the fibres is greater than their length l. Therefore hydrodynamic interactions between the fibres are rare. Coming to the semi-concentrated region, the spacing between the fibres is less than the fibre length l but higher as the diameter d. Because of that hydrodynamic interactions become more frequent. In the concentrated region at least the spacing is in the range of d and non-hydrodynamic effects become more important.

2.1.3 Micromechanics of fully aligned short fibre composites

The theories of micromechanics aim to model a heterogeneous material by an anisotropic continuum. Most theories consider idealised inclusions in an isotropic matrix. For aramid fibres this fits conditionally, because they have microfibrillar structure and are themselves anisotropic. The common theory for short fibre reinforced composites are the Halpin-Tsai [17] equations for transversely isotropic composites. The following assumptions are made. The fibres are uniform, discontinuous, cylindrical, and transversely isotropic. The applied coordinate system is displayed in figure 2.1: The elastic modulus of the composite E_L in

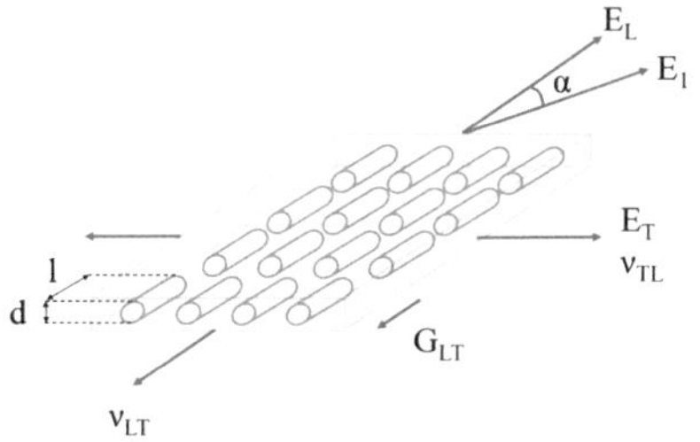

Figure 2.1: Coordinate system for fibre orientation

fibre direction (longitudinal) is defined as:

$$E_L = E_m \left[\frac{1 + 2(l/d)\eta_L \phi_f}{1 - \eta_L \phi_f} \right] \tag{2.5}$$

The elastic modulus of the composite E_T in perpendicular to fibre orientation (transverse) is independent of the fibre aspect ratio:

$$E_T = E_m \left[\frac{1 + 2\eta_T \phi_f}{1 - \eta_L \phi_f} \right] \tag{2.6}$$

$$\nu_{LT} = \phi_f \nu_f + (1 - \phi_f)\nu_m \tag{2.7}$$

$$\nu_{TL} = \nu_{LT} \frac{E_T}{E_L} \tag{2.8}$$

$$G_{LT} = G_m \frac{1 + \eta_G \phi_f}{1 - \eta_G \phi_f} \tag{2.9}$$

$$G = \frac{E}{2 + 2\nu} \tag{2.10}$$

where η_L, η_T and η_G are defined as:

$$\eta_L = \frac{(E_f/E_m) - 1}{(E_f/E_m) + 2(l/d)} \tag{2.11}$$

$$\eta_T = \frac{(E_f/E_m) - 1}{(E_f/E_m) + 2} \tag{2.12}$$

$$\eta_G = \frac{(G_f/G_m) - 1}{(G_f/G_m) + 1} \tag{2.13}$$

where E_m and E_f are the elastic modulus of the matrix and the fibre respectively, while ν_f and ν_m are the Poisson's ratios of the fibre and the matrix. G_m and G_f are the shear modulus of the matrix and the fibre. ϕ_f is the fibre volume fraction and l/d is the fibre aspect ratio.

2.1.4 Effect of fibre orientation

Composite materials can usually be modelled as orthotropic materials for which the stress-strain relations are:

$$\begin{bmatrix} \epsilon_1 \\ \epsilon_2 \\ \epsilon_3 \\ \gamma_{23} \\ \gamma_{31} \\ \gamma_{12} \end{bmatrix} = \begin{bmatrix} S_{11} & S_{12} & S_{13} & 0 & 0 & 0 \\ S_{12} & S_{22} & S_{23} & 0 & 0 & 0 \\ S_{13} & S_{23} & S_{33} & 0 & 0 & 0 \\ 0 & 0 & 0 & S_{44} & 0 & 0 \\ 0 & 0 & 0 & 0 & S_{55} & 0 \\ 0 & 0 & 0 & 0 & 0 & S_{66} \end{bmatrix} \begin{bmatrix} \sigma_1 \\ \sigma_2 \\ \sigma_3 \\ \tau_{23} \\ \tau_{31} \\ \tau_{12} \end{bmatrix}. \tag{2.14}$$

where S_{11} is defined as

$$S_{11} = \frac{1}{E_1} \tag{2.15}$$

For a lamina reinforced by fibres oriented in an angle α to the axis of applied stress it can be written:

$$E_1 = \frac{1}{\frac{\cos\alpha^4}{E_L} + \frac{\sin\alpha^4}{E_T} + (\frac{1}{G_{LT}} - 2\frac{\nu_{LT}}{E_L})\cos\alpha^2 \sin\alpha^2} \tag{2.16}$$

For a random fibre orientation the resulting composite elastic modulus and shear modulus can be calculated according to Bert and Kline [18] by the following simplified formulas:

$$E_{random} = \frac{3}{8}E_L + \frac{5}{8}E_T \tag{2.17}$$

$$G_{random} = \frac{1}{8}E_L + \frac{1}{4}E_T \tag{2.18}$$

2.2 Elastomers

2.2.1 Properties

Elastomers are polymers with occasional cross-links forming a 3D network, which show entropy-elastic behaviour at temperatures above their T_g. They do not melt but can swell. They can own high molecular weights M_W up to 2.000.000 g/mol as for natural rubber (NR). Typically the cross-links are originated chemically or physically by using the double bonds in the backbone. The cross-linking process is called vulcanisation and is typically executed with the help of sulphur and activators or peroxides under supply of heat. Also curing with the help of irradiation or metallic oxides is sometimes applied. The high elasticity of the vulcanised elastomers originates from the entropy-elastic behaviour of the cross-linked macromolecules. The coiled polymer chains are oriented if an external force is applied. If this force is released, the polymer chain is going back to the thermodynamically favoured state of higher entropy of a random orientation.

Since in this work NR and ethylene-propylene-diene rubber (EPDM) are used, the chemical structure of these elastomers is presented in figure 2.2 and in the following they are further introduced. NR is natural product harvested in form of latex from trees in plantations typ-

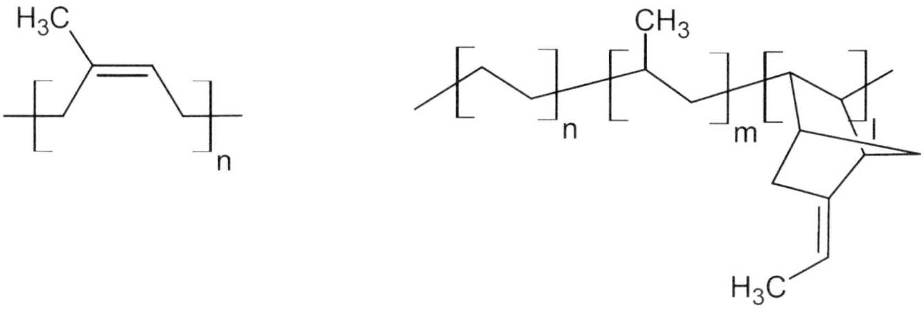

Figure 2.2: Stucture of NR (left) and EPDM (right)

ically in Thailand, Indonesia or Malaysia. Chemically it is cis-polyisoprene, but typical it can contain over 100 other different impurities like proteins, fatty acids, and resins. NR can crystallise under high strain [19]. This strain induced crystallisation increases the modulus and tear energy and therefore inhibits crack propagation. This behaviour is an advantage over synthetic rubbers and makes NR suitable for hard-wearing applications especially in tyres.

EPDM belongs to the group on non-polar rubbers with saturated chain backbone. Therefore it has an excellent resistance against heat, weather or ozone and is typically used in hoses, belts and seals. For EPDM the ethylene content is important, since the propylene and termonomers disrupt the tendency of the ethylene to crystallise.

In general elastomer composites consist of variety of ingredients, typically of the rubber itself, the necessary curatives like sulphur or peroxide, activators, accelerators, inhibitors, zinc oxide, carbon black and silica and processing aids. Typically the proportion of filler in a elastomer composite is given in parts per hundred parts of rubber (phr).

Active fillers with high specific surface roughness like carbon black and silica tend to agglomerate to fractal particle networks and are binding the rubber matrix at their surface in glassy layers [20]. For higher filler volume fractions exceeding a certain threshold they are creating percolation networks. This network leads to an effect described by Payne in 1962 [21]. He described the dramatically decrease of the storage modulus with increasing amplitude of dynamic strain at a constant frequency. The origin of this effect is the break-up of the filler-filler interactions in the 3D filler network (figure 2.3 left). Another effect was described by Mullins et al. [22] for cyclic tensile testing of the same rubber specimen (figure 2.3 right). It is a softening effect of the stress-strain curve which occurs if the load increases beyond its prior maximum value. The hysteresis of the unfilled NR gum is hereby originating from the strain induced crystallisation of the material.

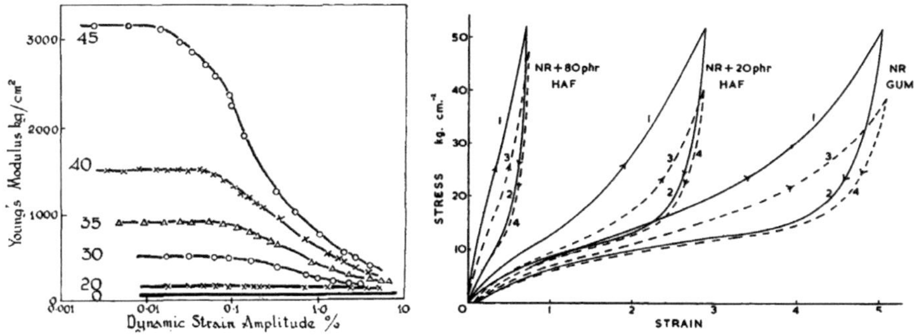

Figure 2.3: Payne effect (left [21]) and Mullins effect (right [22]) of NR vulcanisates with different carbon black content

2.2.2 Elastomers for short fibre composites

In this section a literature review of elastomeric matrices and short fibres used for short fibre reinforced elastomer composites in the past is given.

The elastomeric matrices used for short fibre reinforcement cover the commonly used rubbers like natural rubber (NR), synthetic rubbers like ethylene-propylene-diene rubber (EPDM), isoprene rubber (IR), styrene-butadiene rubber (SBR), acrylonitrile-co-butadiene rubber (NBR), polychloroprene rubber (CR), and thermoplastic elastomers (TPEs) based on polyester, polyamide or polyurethane.

Ibarra [23] reported the dynamic properties of EPDM filled with treated PET fibres by varying the dynamic stress amplitude, fibre content and surface treatment of the fibre. EPDM reinforced with melamine fibres with several dry bonding systems was investigated by Rajeev and co-workers [24]. Manchado and Arroyo [25] compared the mechanical and curing properties of compounds containing short aramid, glass and cellulose fibres in EPDM, NR and SBR. The highest reinforcement was found for aramid fibres.

For NR Datta et al. [26] showed the improvement of hysteresis, heat build-up and cut/-chip/chunk resistance by adding 1–2 phr of sulphur modified aramid fibres to a truck tyre tread compound based on NR. Also improved hysteresis and steel cord adhesion of a NR based compound through the use of 1,5 phr chemically activated aramid fibres was claimed [27]. Murthy and co-workers [28] examined the failure surfaces of short glass fibre reinforced NR composites by scanning electron microscopy. They studied tensile, tear, flexing and abrasion properties of the composites with and without carbon black. Also Cataldo [29] used NR as matrix and reinforced it with carbon fibres. An anisotropic mechanical reinforcement especially for low elongations, reduction of mechanical hysteresis and compression set was reported. Geethamma [30] et al. improved the interfacial adhesion between 10 mm long coir fibres and NR by treatment of the coir fibres with alkali and NR solution. Green strength, tensile modulus, tensile strength and tear strength of the composites were analysed to evaluate the performance of the composites. They also investigated the extent of fibre orientation and the fibre-rubber interfacial strength by solvent swelling studies and scanning electron microscopy. Varghese and co-workers [31] researched the solvent swelling of NR composites reinforced with untreated and acetylated short sisal fibres with a series of alkenes. They found the solvent swelling was reduced by increasing fibre content and adding bonding agent. Murty et al. [32] explored the viscoelastic properties of short jute fibre reinforced NR compounds as a function of temperature and at different levels of adhesion. Finally, Bipinbal [33] and Kutty compared the incorporation of short nylon fibres in NR by latex masterbatch and conventional method.

Agrawal et al. [34] studied reinforcement of short aramid fibre in comparison with carbon black and silica in NBR. Mechanical properties, especially tear strength of the composites was determined and compared with particulate filled rubber vulcanisates with respect to the surface treatment of the fibres. Also Rajesh et al. [35] investigated NBR composites filled with short nylon fibres of various lengths and fibre contents. The influence of different rubber cross-linking systems on the properties of the composites was examined. They found a fibre length of 6 mm and peroxide curing to be optimum for the best balance of properties. Swollen composites vulcanised with the peroxide system were found to have higher rubber volume fraction than that in the sulphur system, which indicated better rubber-fibre interaction. Chakraborty and co-workers [36] researched short jute fibre reinforcement of carboxylated nitrile rubber (XNBR). The effect of the accelerator system, fibre concentration, the role of silica in the bonding system, different bonding systems, aging resistance and rubber filler interaction as well as processing characteristics were investigated.

For CR Wada et al. [37] studied the orientation of short aramid fibres in CR composites by using soft x-ray photographs. Setua and Dutta [38] investigated the role of bonding system for CR filled with silk fibre. They studied three different dry bonding systems. For the best dry bonding system, scanning electron microscopy studies of tensile, tear, abrasion, and flex failed surfaces of both unfilled and fibre filled were executed. Finally, Lee [39] reported improved mechanical properties of CR filled with short glass fibres depending on their orientation.

The use of short aramid fibre in silicone rubber composites and its processing by liquid injection moulding was investigated by Orlob and Reußmann [40]. They tried to increase the known insufficient adhesion between aramid fibres and silicone rubber by mechanical fibre surface roughening and various chemical coupling agents.

The effects of fibre length, concentration and shear-rate/stress on melt viscosity behaviour of untreated and treated short sisal fibre reinforced SBR composites was investigated by Kumar and Thomas [41]. They also studied the dependence of melt viscosity on temperature, flow behaviour index, melt elasticity, extrudate distortion. Wangenheim and Schuster [42] investigated the influence of various nano fibres in SBR and compared the mechanical properties and dispersion with SBR compounds filled with carbon nano tubes or carbon black. Praveen et al. [43] studied the effect of nanoclay and short aramid fibres on the mechanical and damping properties in carbon black filled SBR compounds. They found that nanoclay had a negative effect on the properties of the composites. Just the elongation at break found was higher for the samples with nanoclay.

Beside matrices based on side rubbers also rubber blends and thermoplastic elastomers were in the focus on research of short fibre reinforced elastomers.

Datta [44] studied the effect of low concentrations of untreated and treated 3 mm fibres in truck compounds containing NR and BR and achieved reduced hysteresis. Also Kashani [45] investigated the fibre orientation, mechanical and dynamical properties as well as heat build-up and abrasion on NR/BR compounds. Reduced rolling resistance and heat build-up was reported for a NR/BR blend reinforced by low contents of different oriented various short fibres by Rijpkema [46].

Gunnarsson and Nyman [47] investigated the influence of various fibres as reinforcement in a NR/SBR blend regarding the tensile properties, tear strength and abrasion resistance. Ahmad and co-workers [48] studied graft-copoly (ethylene/maleic anhydride) (PE-g-MA) as a compatibiliser between aramid fibres and thermoplastic natural rubber (TPNR) composites. The interaction between the matrix and fibre was improved with the addition of PE-g-MA. The loss factor tan δ peak decreased with the higher fibre volume fraction in the system with PE-g-MA and increased in the system without PE-g-MA. In another work Ahmad et al. [49] studied the effects of chemical treatment on short aramid fibres in blends of polyethylene and NR by using micromechanical and mechanical testing. They investigated untreated fibres and treated fibres with epoxy adhesion activation. Introducing a resorcinol formaldehyde latex (RFL) treatment improved the interaction between the aramid fibre and ENR matrix.

Arroyo and co-workers [50] researched the effect of addition of EPDM and short aramid fibres on polypropylene crystallisation kinetics and tensile behaviour. Aramid fibres and EPDM were found to be effective nucleating agents for PP crystallisation.

In case of thermoplastic polyurethane (TPU) Kutty and Nando [51] explored the influence of fibre content and orientation to the mechanical and dynamic mechanical behaviour of short aramid fibre filled TPU. They found increased strength of the composite above 10 phr of fibre content. Tan δ was reduced progressively with growing fibre volume fraction. Impact strength was reduced considerably at all fibre contents irrespective of fibre orientation. Scanning electron microscopy showed the existence of a kinking stage, which causes severe fibre breakage during processing.

Also Vajrasthira and co-workers [52] investigated the mechanical and dynamic mechanical properties of TPU reinforced with two types of aramid short fibres. Both types of fibre reinforced TPU effectively without any surface treatment. The broadening of the tan δ peak in dynamic mechanical analysis revealed a good fibre-matrix interaction. Akbarian [53] also studied the mechanical and dynamic mechanical behaviour of short aramid fibre reinforced TPU. In addition they investigated the rheology and found a power law behaviour for all composites and increased viscosity with fibre content. Scanning electron microscopy showed extensive peeling and fibrillation of the fibres in the compound.

For ethylene octene copolymer (EOC) Naskar und Shibulal [54, 55] investigated the effect of fibre content and initial length of aramid fibres. They used maleic anhydride grafted polybutadiene as dispersion and coupling agent and achieved a good balance of properties for 6 mm long fibres.

2.3 Fibres

A wide variety of fibres has been studied to reinforce elastomers. Natural fibres, mainly cellulose [16, 56, 57], sisal [31, 41, 58], jute [32, 36], silk [38, 59], coir [30] and oil palm [60] based fibres, have the advantage that they are renewable and sometimes show a better bonding to rubber. Their complicated fabrication and fluctuating reinforcement levels are their disadvantages.

Compared to natural fibres synthetic and inorganic fibres like nylon [35, 61, 62], polyester [23, 46, 63], glass [16, 32, 39, 61, 64], carbon [29, 65] and aramid [16, 25, 43, 45–47, 53–55, 61, 66] exhibit continuous fabrication and therefore better tolerances in shape and uniformity. Brittle fibres like carbon and glass are not so common for reinforcing elastomers, because of dramatic fibre breakage during processing.

2.3.1 Aramid fibres

Aramid belongs to the group of liquid crystal polymers (LCP). Three main types are commercially available: M-aramid poly(m-phenylene isophtalamide) (MPIA) is sold under the tradenames Nomex or Conex. P-aramid poly(p-phenylene terephtalamide) (PPTA) is sold under Tradenames Kevlar or Twaron. Co-poly-(paraphenylene/3,4'-oxydiphenylene terephthalamide) (PP/ODPTA) is sold unter the tradename Technora. The mechanical properties

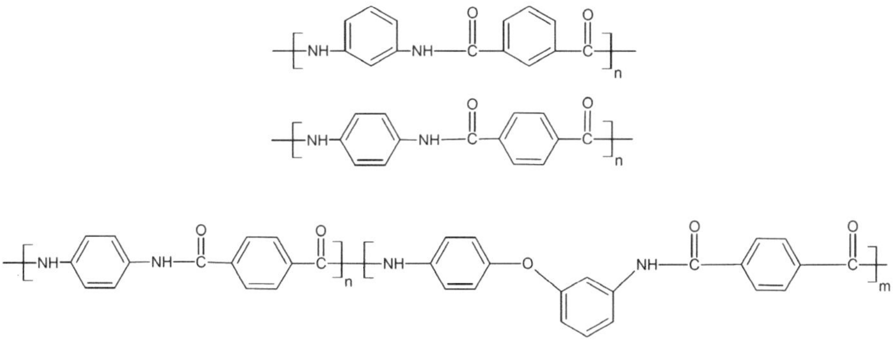

Figure 2.4: From top to bottom: M-aramid poly(m-phenylene isophtalamide) (MPIA), P-aramid poly(p-phenylene terephtalamide) (PPTA) and Co-poly-(paraphenylene/3,4'-oxydiphenylene terephthalamide) (PP/ODPTA)

of these fibre types can be obtained in table 2.1. Since the main focus in this work is on p-

Fibre type	Young's Modulus [GPa]	Tensile strength [MPa]	Elongation at break [%]
PPTA	24 – 25	2400 – 3000	3 – 4
MPIA	8 – 10	600 – 700	34 – 45
PP/ODPTA	20 – 21	3000 – 3250	5 – 7

Table 2.1: Mechanical properties of different types of aramid fibre [44]

aramid (PPTA) this subsection will be mainly about this aramid type and its development, fabrication, properties, structure and failure modes.

2.3.1.1 Development and production

PPTA was developed in the late 1960s. DuPont and Akzo Nobel were both involved in development and forced the commercialisation. Teijin later bought the aramid part of the fibre division from Akzo Nobel. This is the reason why p-aramid fibres from DuPont are sold under the trademark Kevlar and from Teijin under trademark Twaron. P-aramid can be prepared various routes. Classically it is produced by condensation polymerisation of p-phenylene diamine with terephthaloyl chloride in a suitable solvent such as hexamethylphosphoramide or N-methylpyrrolidone. The HCl formed during polymerisation is neutralised by washing the resulting polymer with dilute NaOH [67]. Through the dry-jet wet spinning process in a coagulation bath the macromolecules are oriented in direction of flow. Because of the fact that the cooling rate of the filaments is faster at the outside, the outside skin is less crystalline, which leads to the typical macroscopic structure [68].

2.3.1.2 Structure

P-aramid fibres are containing several levels of superimposed microscopic and macroscopic structures. They include the crystalline structure, pleat structure, fibrillar structure and skin-core structure, which are shown in Figure 2.5 . The crystalline structure of a PPTA

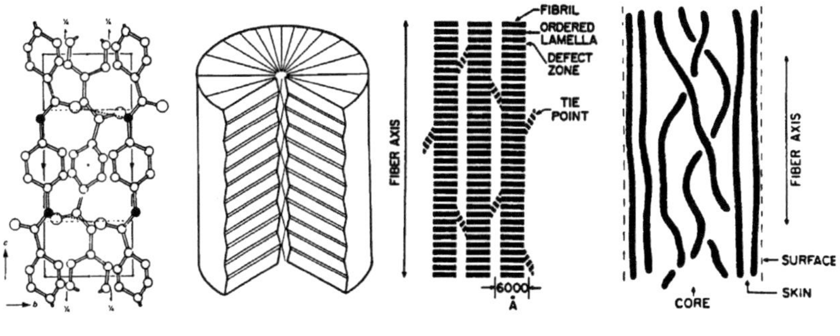

Figure 2.5: From the left to the right: shematic diagrams of crystalline structure [69], pleat structure [70], fibrillar structure [71] and skin-core structure [71]

macromolecule brings some unique properties. It is a linear organic polymer with high stiffness and covalent bonds along the fibre axis with high packing density. The molecules are connected with hydrogen bonds while the aromatic rings causing the stiffness and uniform orientation.

Dobb et al. [72] found transverse bands of p-aramid fibre by electron microscope dark-field image technique. It was found that there is 250 nm spacing between these bands. A structural model of radial pleated sheets was developed based on those observations. The transversal bands are arranged at equal but opposite angles of about 170°–175°. Between these bands a narrow transition region exists, where the PPTA-macromolecules are parallel to the plane of the fibre section. The pleated sheet structure is also morphologically superimposed on the fibrillar structure.

Because of the highly ordered structure p-aramid fibres fibrillate upon abrasion. The reason for that are the lack of lateral forces, except for the van der Waals and the hydrogen bonds forces. According to Panar et al. [71] the fibrils are thereby about 600 nm wide, up to several centimetres long and are oriented along the fibre axis. These fibrils are joined by bundles of tie fibrils and show the crystalline defect layers at 35 nm spacing. A skin-core differential in density, voidage and fibrillar orientation of p-aramid fibres caused by the production process lead Panar et al. to a simple skin-core model. In this model the surface fibrils are uniformly axially orientated, where the core fibrils are imperfectly packed and ordered. P-aramid gives diffuse scatter in small angle X-ray diffraction (SAXS), which suggests the presence of microvoids in fibre core, elongated with their long axes parallel to the fibre axis.

2.3.1.3 Properties

PPTA macromolecules are rod-like with a reported molecular weight M_W up to 40.000 g/mol [73]. PPTA fibres shows outstanding strength-to-weight properties with a density of 1,44 g/cm^3 and a filament diameter of 12–15 μm [74, 75]. They also exhibit a high thermal stability (T_G at 370°C, degradation starts from 500°C [76]), are nonconductive, sensitive to compression and radiation from ultraviolet light [77, 78] and radioactive sources [79]. Compression of aramid fibres composites generally does not result in complete fibre failure, but rather a sequence of buckling and fibrillation takes place. After heavy buckling, the fibres easily fail in tension with short split lengths as a result of their compression-induced PPTA macromolecular chain scission [68]. The long-time exposure to occurring fibrous dust was investigated by several studies [80, 81]. The PPTA fibre is hygroscopic and can absorb up to 3,5–4,5 $wt\%$ moisture in ambient temperature in 100% relatively humidity [78].

2.3.1.4 Failure modes

Aramid fibres easily buckle and form kink bands at 50° to 60° [82–84] when compression or torsion is applied to them in fibre direction. Deteresa et al. [83] reported that after applying 3% of compressive strain the fibres lost 10% of their tensile strength. They found longitudinal fibre splitting initiated at the fibre surface, where the shear strain had a maximum. This effect is caused by the relatively weak van der Waals forces and lateral hydrogen bonds of aramid fibres together with the presence of residual impurities in the interfibrillar regions.

Similar to the findings for compression or torsion the fibres will fail in tension by axially splitting to 600 nm thick fibrils reported by Panar et al. [71], which showed lengths between 500 and 800 μm [75]. The failure will be initiated by any discontinuity on the fibre skin and cracks will propagate along the fibre axis causing detachment of fibrils.

Hashmito and co-workers [85] investigated the degree of fibrillation as function of processing time in a vibrating ball mill. They found that with increasing fibrillation the interaction between the fibres and a powder matrix was increased by mechanical interlocking. Similar findings were reported by Roebroeks [86].

2.3.1.5 Applications

PPTA is applied in form of yarns, staple and short fibres, powder, pulp, fabrics and laminates. Aramid is typically used in cables, ropes, composites, coated fabrics, tapes, brake pads, webbings, bulletproof vests and tyres.

2.4 Fibre-elastomer interaction

In this section an overview is given of the possibilities to increase the fibre-elastomer interaction and procedure to characterise the physical interaction with the help of contact angle measurement. Fibre-matrix adhesion is mainly affected by different mechanisms: the amount of interface surface area and the interaction between fibre and matrix.

2.4.1 Possibilites to increase fibre-elastomer interaction

Although the chemical interaction between aramid fibres and polymeric matrices might be low, the constriction of the rubber matrix under tension increases the mechanical friction at the interface by rising the normal forces [47]. However, this effect can be enhanced by several methods.

The first option to increase the fibre-matrix interaction is to increase the amount of interface surface area. Roebroeks [86] stated that controlled surface fibrillation improved the static transverse strength of aramid reinforced composites considerably and that it was compatible with all matrices in contrast to dips. Also Orlob and Reußmann [40] showed that the controlled fibrillation as well as higher initial fibre length improved the mechanical properties at low elongations. The controlled surface fibrillation is already commercially available and is called pulp [74, 78]. The highly fibrillated surface of these type acts as an anchor in the matrix and creates therefore a strong mechanical bond to the matrix. Aramid pulp is fabricated from chopped filament yarn by shearing a fibre suspension in water in a refining process [7]. A disadvantage of pulp is the difficult dispersion of the fibres in the matrix, which can be solved by preparation of a masterbatch with carbon black and elastomer [78, 87].

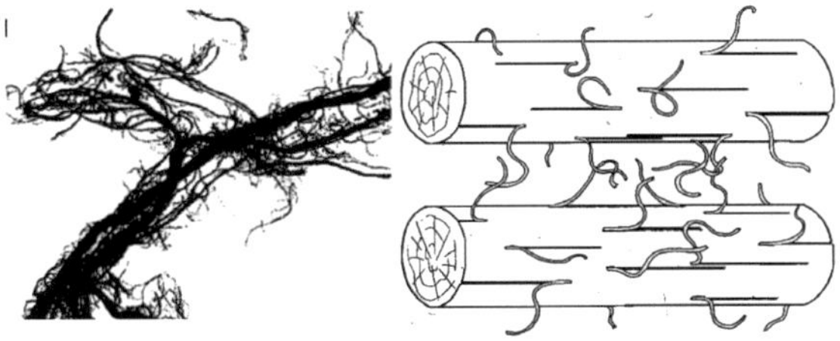

Figure 2.6: Aramid in pulp form [68], scheme of controlled fibrillation [86]

The second option is to dip the fibres to enhance the interaction with elastomer matrices. The high crystallinity and steric hindrance of the bulky aromatic rings make it difficult to bond the aramid fibres to the elastomer matrix. Before any additional treatment the aramid fibres get a predip of an 'oily' finish component for processability reasons [68, 88]. Varisthira and coworkers [52] studied the reinforcing properties of m-aramid (Conex) and copoly-p-aramid (Technora) without any additional adhesion treatment in TPU composites. The morphologies of cryogenically fractured surfaces of the composites and extracted fibres, investigated with scanning electron microscopy, revealed possible polar-polar interactions between the aramid fibres and the TPU matrices. However, because this treatment is not sufficient to enhance the adhesion to rubber, several other treatments are applied. A typical chemical treatment of aramid fibre types for improvement of adhesion to epoxy and rubber matrices is the application of an epoxy-based finish to their surface, and subsequent curing thereof. De Lange et al. [89, 90] investigated the effect of oily finish components on the interface strength of aramid-epoxy composites. Therefore Twaron yarn was treated in a one-step finish process with combinations of adhesion improving epoxy-amine components and processability improving oily components. They studied the macroscopic composite properties as a function of the amount of oil and confirmed them by micro-mechanical adhesion results from single filament pull out measurements and micro Raman spectroscopy. Differences in the amount and distribution of epoxy, fixed to the yarn surface, shown by X-ray photoelectron spectroscopy (XPS), contact angle measurements and atomic force microscopy confirmed the results of macroscopic level. Finally, they confirmed improved adhesion of the activated Twaron yarns under the presence of small amounts of oil.

Also resorcinol formaldehyde latex (RFL) treatment is often applied to increase adhesion. The RFL dip is an emulsion of a rubber latex in a solution of resorcinol and formaldehyde in water. An RFL dip is prepared in two stages. In the first step, an aqueous solution of resorcinol and formaldehyde is matured for several hours at room temperature. Sodium hydroxide is added so that the mixture becomes basic. During the maturation process, some degree of condensation takes place. In the second step, the resin solution is added to a mixture of latex and water. Wennekes [91, 92] studied the influence of rubber curatives in the rubber compound on the adhesion to RFL-treated aramid cords. He used scanning electron microscope coupled to an energy dispersive X-ray spectrometer to analyse the interfaces between rubber compound and RFL-dip. Also the influence of the type of latex used in the RFL formulation, on the adhesion to several rubber compounds was studied. De Lange

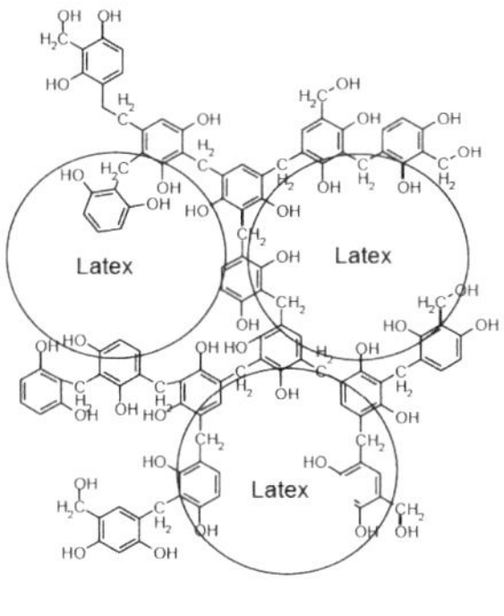

Figure 2.7: Schematic RFL morphology [91]

et al. [88] concluded that the adhesion of RFL dipped Twaron yarns to natural rubber is dependent on the amount of oil used influencing the amount of epoxy-amine fixed to aramid surface. Ahmad et al. [49] researched the effects of chemical treatments on aramid fibres in Epoxidized Natural Rubber. They investigated untreated fibres and treated fibres with epoxy adhesion activation. Introducing RFL treatment improved the interaction between the aramid fibre and ENR matrix, which was confirmed by micromechanical and mechanical testing. Also compatibilisers were investigated to enhance the interface between aramid fibres and elastomers. Graft-copoly(ethylene/maleic anhydride) (PE-g-MA) was studied [48] as a compatibiliser between aramid fibres and thermoplastic natural rubber (TPNR) composites. The interaction between the matrix and fibre improved with the addition of PE-g-MA.

The peak of the loss factor decreased for higher fibre contents by the presence of PE-g-MA. The Sartomer company [93] reported that maleated low molecular weight polybutadienes (PBDMA) improved the physical properties of cured rubbers through increase of fibre orientation, network density of the matrix, and adhesion between fibre and the rubber matrix. Results showed that PBDMA acted as an effective wetting agent, reactive compatibiliser and coagent and improved the miscibility of chopped aramid fibres with rubbers. Also Naskar and Shibulal [54, 55] applied a maleic anhydride grafted polybutadiene as dispersion and coupling agent of RFL dipped aramid fibres in ethylene octene copolymer (EOC) and got improved dispersion and fibre-matrix interaction.

Inagaki et al. [94] investigated the surface modification of PPTA films with 2-aminoethanethiol to increase the adhesion to silicon rubber. XPS showed that oxygen functionalities were generated and adhesion was improved. Also Orlob and Reußmann [40] tried to improve the adhesion between PPTA short fibres and silicone rubber by treatment of the fibres with vinyltriethoxysilane. The treatment caused dispersion problems and the interaction between the fibres and matrix was not increased. With the help of isocyanatosilane and epoxysilane the fibre-matrix adhesion was finally improved.

Another chance to increase the fibre adhesion of PPTA fibres to various matrices is plasma- or irradiation treatment. Effects of various fibre surface treatments on the interfacial micromechanics of aramid fibres in an engineering thermoplastic elastomer (polyether amide block co-polymer) were investigated by Coffey and co-workers [95]. The distribution of interfacial shear stress along fibres in single-fibre model composites was determined using Raman spectroscopy. Chloride end-groups were grafted to the aramid by methacryloyl chloride. Also plasma treatment with argon, oxygen, nitrogen and ammonia gas plasma treatment was carried out. N-vinyl-2-pyrollidinone was grafted onto the aramid fibre surface using a photo-polymerising procedure. Also coupling agents like Neopentyl(diallyl)oxy, tri(N-ethy1enediamino)ethyl titanate and Neopentyl (diallyl)oxy, tri(m-amino)phenyl zirconate were investigated for better interface properties of rubber to matrix. They found that fibres treated with methacryloyl chloride and N-vinyl-2-pyrollidinone showed effective stress transfer.

Hassan et. al [96] investigated the use of reclaimed rubber powder (RPP) by the addition of maleic anhydride mixed with NR and glass fibres in an open two-roll mixing mill. These composites were irradiated with 30 and 50 kGy irradiation doses. Results showed that the tensile strength and swelling resistance increase with increasing RRP content in the NR/RRP blends, whereas the elongation at break exhibit opposite trend. Watanabe et. al [79] applied a low temperature plasma and excimer laser irradiation method to aramid fibres to enhance the adhesion to rubber matrices. It was shown that the adhesion could be enhanced, if immediately after low temperature plasma the fibres were treated with RFL. This effect was more pronounced for PPTA than for PP/ODPTA. Also de Lange [97] studied plasma treatment to increase adhesion of PPTA in rubber and epoxy matrices. He concluded that the adhesion performance of plasma treated p-aramid yarn was lower than that of chemically treated ones.

Nitrogen plasma and electron-beam irradiation were studied [98] to create surface radicals and initiate grafting. The surface grafting was more important when the PPTA was irradiated with a cold plasma and if water was used as the solvent. When the PPTA was electron beam irradiated the polymer crystallinity degree reduced the grafting even with a high radical concentration. Sharma [99] from The General Tire and Rubber Company registered a patent on cold gas plasma treatment of various gases on p-aramid cords before dipping it with chemical treatments. It was shown that the cord rubber adhesion could be significant improved.

A comprehensive review of the possible ways to increase fibre-matrix adhesion of reinforced rubber composites containing short aramid fibres is given by Rajeev [11].

2.4.2 Predictor surface energy

That the surface energy and polarity calculated from contact angle measurements is a suitable method to quantify the degree of physical interaction between fillers and elastomers was shown by Stöckelhuber et al. [100]. Young [102] found that the wetting of solids and

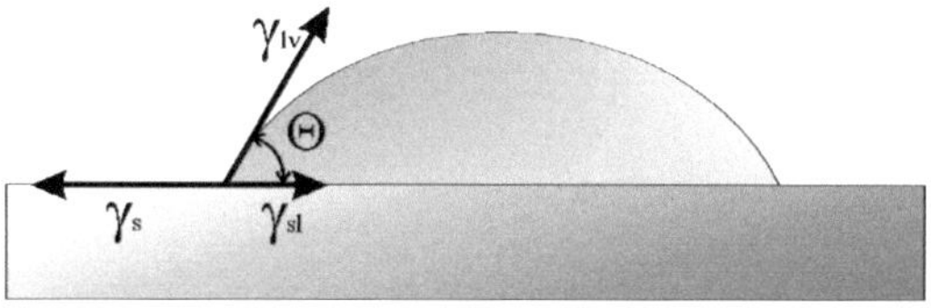

Figure 2.8: Physical forces determining the three phase contact of liquid droplet [101]

fluids is determined by the interfacial tensions of the components according to the following equation:

$$\gamma_{lv}\, cos\, \theta = \gamma_s - \gamma_{sl} \tag{2.19}$$

where γ_s is the surface tension between solid and liquid, γ_{sl} the interfacial tension between solid and liquid, γ_{lv} is surface tension between liquid and its vapor, while θ is the contact angle. The total surface energy γ can be splitted in a dispersive γ^D and a polar part γ^P.

$$\gamma = \gamma^D + \gamma^P \tag{2.20}$$

The work of adhesion between a solid and a liquid phase was defined by Dupré [103] as:

$$W_a = \gamma_s + \gamma_l - \gamma_{sl} \tag{2.21}$$

It can be interpreted as the work, which has to be applied to separate two phases, which are already in contact. By combination of equation 2.21 with Youngs equation 2.19, the Young-Dupré equation for the work of adhesion is given by:

$$W_a = \gamma_l(1 + cos\theta) \tag{2.22}$$

Fowkes [104] stated that the work of adhesion W_a between two materials is defined as:

$$W_a = 2\sqrt{\gamma_1^D \gamma_1^D} + 2\sqrt{\gamma_s^P \gamma_l^P} \tag{2.23}$$

The Young-Dupré equation can then be reformulated by insertion of the equation by Fowkes to:

$$cos\, \theta = -1 + \frac{2\sqrt{\gamma_s^D \gamma_l^D}}{\gamma_l} + \frac{2\sqrt{\gamma_s^P \gamma_l^P}}{\gamma_l}) \tag{2.24}$$

The free energy of immersion ΔG_i is defined as the difference of the interfacial energy between the solid filler surface and the circumfluent polymer,γ_{sl}, and the surface energy of the filler particle,γ_s.

$$\Delta G_i = \gamma_{sl} - \gamma_s \tag{2.25}$$

The free energy of immersion is a predictor for the contribution of thermodynamics to the dispersibility. If ΔG_i has a negative value, then the wetting of the filler by the rubber is thermodynamically favoured. Positive values of ΔG_i show that an immersion or dispersion of fillers is not a preferred thermodynamic state for this filler-rubber system. The direct measurement of the interfacial energy γ_{sl} is not possible. In combination with Youngs equation the free energy of immersion can be written as:

$$\Delta G_i = -\gamma_l \cos\theta \tag{2.26}$$

Due to the fact that the contact angles of rubber polymers on a filler surfaces are not accessible for direct measurements, the previous equation can be further modified by insertion of Fowkes equation 2.23, which leads to:

$$\Delta G_i = \gamma_l - 2(\sqrt{\gamma_s^D \gamma_l^D} + \sqrt{\gamma_s^P \gamma_l^P}) \tag{2.27}$$

Finally, the work of adhesion can be rewritten as:

$$W_a = 2(\sqrt{\gamma_s^D \gamma_l^D} + \sqrt{\gamma_s^P \gamma_l^P}) \tag{2.28}$$

It can be used to predict the work which has to be applied to separate two phases which are already in contact.

2.5 Compounding and Processing

2.5.1 Rheology

The rheological behaviour of elastomer composites is important for dispersion, fibre breakage and orientation. It is known that the addition of small particulates like carbon black enhances the magnitude of viscosity, especially for low rates of deformation [105]. For highly carbon black filled compounds yield stress values were found, for which there was no flow observable anymore. However, compounds with larger particles, for examples aramid fibres did not exhibit this behaviour.

The viscosity of rubbers is determined accordingly to standard DIN 53523 [106] with a special shearing disc viscometer and the gained Mooney viscosity is defined as the shearing torque resisting rotation of a rotor surrounded by rubber in a cylindrical cavity. The Mooney scorch is determined at temperatures at which vulcanisation may occur measuring the initial rate of vulcanisation due to the torque increase.

The influence of various short fibres on Mooney scorch and Mooney viscosity was investigated by Foldi [61]. Mooney viscosity (ML 1+4 at 100°C) increased with increasing fibre content. For 10 phr fibre content the change was less than 10% and for 20 phr it was less than 50%. Contrary to that the investigated fibre lengths between 5 and 64 mm had very little effect on Mooney viscosity and Mooney scorch. Also Rajeev et al. [24] reported that Mooney viscosity increased by raising fibre content. The variation of Mooney viscosity with volume fraction of fibre ϕ_f followed the equation:

$$\frac{M_f}{M_g} = 1.060\left[1 - 0.02((l/d)\phi_f - (l/d)^2\phi_f^2)\right] \tag{2.29}$$

where M_f and M_g stand for the Mooney viscosity of the fibre filled compounds and gum compound, respectively.

Kumar et al. [41] studied the effects of fibre breakage, length, concentration and shear-rate/stress on melt viscosity behaviour of untreated and treated short sisal fibre reinforced SBR composites by using an Instron capillary rheometer at 120°C. The effect of fibre length on the melt viscosity of SBR composites was studied for a series of mixes containing 20 phr short sisal fibres with a length 2, 6 or 10 mm, respectively. At low shear rates, the short fibres increased the viscosity much more than at high shear rates, which is known as pseudoplastic behaviour. The explanation was that fibres having shorter length can be more easily aligned and dispersed than those having higher lengths along the direction of flow. The same trend was found for the variation of fibre content from 5 to 20 phr. With increasing fibre content the melt viscosity was raising. For increasing shear stresses the melt viscosity reduced, because the polymer molecules as well as the fibres oriented in the direction of shear. Also die swell measurements were carried out to understand the elastic effects during extrusion. The extruded samples were analysed with help of optical and electron microscopy to characterise the surface morphology and the extrudate deformation. The extrudate distortion and die swell behaviour of the composites were reduced by increasing short sisal fibre content.

2.5.2 Dispersion and fibre breakage during mixing

Goettler et al. [107] divided mixing in two distinct actions: dispersive and distributive mixing. Dispersive mixing is an action, which lowers the agglomerates size. Opposite to that distributive mixing is an action which enhances the randomness of the filler distribution in the matrix without reducing the filler size further. The main goal is to disperse the fibres in the elastomeric matrix for optimal performance, since agglomerates act as initiators for cracks. High shear forces are occurring during the mixing and support the dispersion process. Contrary to that, these high shear forces will shorten the fibre length dramatically and lower the performance of the composite.

In this context O'Connor [16] studied the mixing of nylon, cellulose, glass, carbon and aramid fibres into nitrile rubber and natural rubber on a two roll mill. He compared the mixing times for sufficient dispersion and the fibre lengths and aspect ratios before and after mixing. As expected the brittle fibres like glass and carbon gave relatively poor performance due dramatically high fibre breakage during mixing. Aramid, nylon and cellulose, maintained aspect ratios sufficient for good mechanical properties. Similar trends for these fibre types were confirmed by several authors. For example Kumar and Thomas [58] found that sisal fibres with 6 mm initial length obtain only small degree of fibre breakage by two roll mixing. The length reduction of 6 mm silk short fibres incorporated in NR and PE blends was investigated by Akthar and coworkers [59]. They received fibres with series of bends and kinks and a fibre length of 4,5 mm after mixing. Lee [39] reported that the length of glass fibres reduced from 12,8 mm down to 268 μm during the processing in CR. Ashida [108] showed similar trends in figure 2.9 for fibres with 6 mm initial length. I: polyethylene terephthalate (PET), II: nylon 6, aramid and rayon, and III: carbon and glass. To avoid fibre breakage

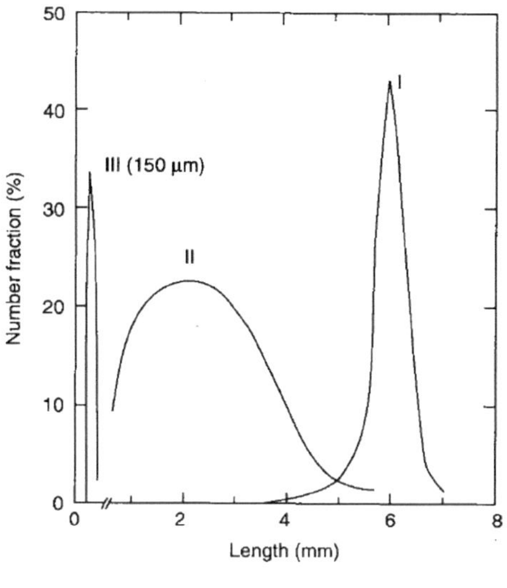

Figure 2.9: Fibre breakage for different fibre types with 6 mm initial length [9]

different methods were used to get the best compromise between adequate dispersion and sufficient fibre length. Ortlepp and co-workers [109] investigated the use of defibrated fibres for an equal incorporation and homogeneous arrangement in various thermoplastics and elastomers. A modified hammer mill was used to open the short cut fibres bundles before mixing in.

A cement method was used by Jana et al. [65, 110] for 6 mm long short carbon fibres in polychloroprene rubber (CR). A paste of masticated CR and chloroform was created. Then the carbon fibres where dispersed in that paste by stirring and after drying sheets were formed on a two roll mill. The resulting fibre lengths could be kept at a high level compared to conventional mixing, which resulted in increased hardness and modulus. Goettler and Cole [12] used a similar method for cellulose fibres. A masterbatch containing a high concentration of fibres and a polymeric binder was diluted in the rubber matrix. The state of dispersion and breakage of these treated fibres was depending on the rheology of the coating formulation. Other important factors were the individual degree of fibre wet out and separation, the viscoelasticity of the coating material and its compatibility with the composite matrix rubber. Also the mixing conditions, like process temperature, mixing speed, mixing time, mixing sequence, mixer geometry and the degree of chamber volume fill played an important role. For the mixing protocol upside-down and two pass mixing protocols provided improved state of dispersion. The dispersion also improved with increasing unit power input and mixing time. Hereby the fibres dispersed rapidly during the early stages of mixing, but the dispersion gradually slowed down as mixing continued. Since the mixing process involved separation of the individual fibres from the bundles, a minimum force had to be exceeded. This threshold value depended on the matrix viscosity. The increase of the stock viscosity had two beneficial effects: total energy input built up faster with time and higher stresses were generated to disperse the fibres more easily. The presence of the fibres did cause a higher rate of heat generation and therefore it was necessary to reduce the batch size by about 10% and ran the mixer at lower speed to avoid scorching of the batch. For highly filled EPDM hose compounds 75% degree of fill (DOF) of the mixer and for CR v-belts 65% DOF were recommended [107]. Bipinbal [33] and Kutty compared a latex masterbatch method with conventional mixing of short nylon fibres in NR. They claimed that by using the latex masterbatch method the mixing time needed for sufficient dispersion as well as the fibre breakage were reduced.

To compare the dispersion for different elastomer compounds in different sized Banbury mixers, which have different temperature profiles and therefore a variation of stock viscosity, Shen and Rains [111] introduced a dimensionless dispersion number. Machine parameters like rotor length, rotor diameter, rotor tip clearance, mixing chamber volume, rotor speed and mixing time were included in this number. Based on the probable number of passes through the shear zone for an ingredient to be dispersed, it made scale-up possible without the necessity of matching the temperature profile. Therefore it was assumed that the general flow pattern in the mixer was independent of mixer size, that the shear stress generated in shear zone at mixing temperatures was larger than the interaggregate cohesive force of the fibre bundles and that the dispersion rate is independent of the shear rate. The dimensionless dispersion number N_{RS} during a mixing period t at rotor speed N was defined as:

$$N_{RS} = (V_S/V_0)Nt \tag{2.30}$$

where V_S is the shear volume and V_0 is the net chamber volume. V_S is defined as:

$$V_S = \pi Lh(D + h) \tag{2.31}$$

where h is the rotor tip clearance, D is the rotor diameter and L is the rotor length. Based on this parameter the dispersion of short fibres can be predicted from one size of internal mixer to another size mixer as well as the mixing time required for a given dispersion rating.

Fibre dispersion of chopped glass fibre bundles in polystyrene for different processing conditions was investigated in internal mixer by Kuroda et al. [112, 113]. They claimed that real glass fibre bundles were dispersed by a combination of rupture and erosion mechanism. The idealised dispersion of a chopped glass bundle was schematically presented in figure 2.10 for understanding the evaluation of fibre diameter distributions. Key values for the

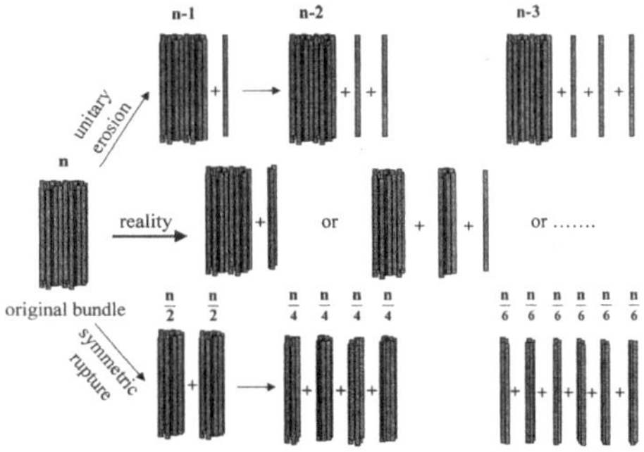

Figure 2.10: Idealisation of dispersion mechanisms for fibre bundles [112]

description of the dispersion are the following. The shear rate $\dot{\gamma}$ is dependent on the rotor speed N, the rotor diameter D and the clearance h between mixer wall and rotor tip:

$$\dot{\gamma} = \frac{\pi D N}{h} \tag{2.32}$$

The total strain Γ is the product of the shear rate $\dot{\gamma}$ and the mixing time t:

$$\Gamma = \dot{\gamma} \cdot t \tag{2.33}$$

The time dependent shear stress $\sigma(t)$ can be obtained from the torque $M(t)$, the rotor diameter D and the surface area of the rotors A:

$$\sigma(t) = \frac{2M(t)}{A \cdot D} \tag{2.34}$$

Finally, the specific energy input e can be calculated from the product of the time dependent integral of the torque $M(t)$ and rotor speed N divided by the batch mass m:

$$e = \frac{1}{m} \int_{t=t_0}^{t=t_E} 2\pi N M(t) \cdot dt \tag{2.35}$$

Flow visualisation of filler dispersion was investigated for carbon black in rubber compounds with the help of glass plate as cover for an internal mixer [114].

2.5.3 Fibre orientation during processing

Moghue [66] studied the effect of milling parameters on the fibre orientation for CR and NBR filled with cellulose and aramid fibres respectively. He concluded that the fibre orientation depends primarily on mill gap and not on roll speed, roll speed ratio or number of passes. The latter would be important only if the stock was carefully folded between the passes. Akhtar and coworkers [59] confirmed this thesis stating that a small nip gap with single pass through the two roll mill gave the best anisotropy for short silk fibre reinforced NR/PE blends. Goettler et al. [115] studied the behaviour of short cellulose fibre reinforced EPDM hose compounds. They stated that the highly pseudoplastic nature of short fibre suspensions is resulting in a blunted velocity profile in channels with uniform cross section, which is caused by the interactions between the long fibres of a high aspect ratio. For higher filled elastomers (> 6 vol%) the applied shear fields align the fibres in flow direction in a thin layer at the wall of the channel. The central part flows as a solid plug. Shear is diminished by limiting the length of the flow path in the die and by heating the die walls. They also reported about the different die geometries and their effect on fibre orientation. In all cases, an expansion of the flow area following a constriction was necessary to turn the fibres from their otherwise axial position. Goettler [56] also moved the alignment of the die parts during extrusion to set a curved shape in the hose right at the die head. Also it was reported that the die swell and the extrudate distortion were reduced. Van der Pol [7] reported that aramid pulp is more feasible for extrusion than dipped chopped aramid fibres. Kashani [45] confirmed adequate orientation for a certain extruder screw speed quantitatively by swelling and scanning electron microscope on fractured surfaces. Clarke and Harris [116] reported of special moulding technique to increase fibre orientation and got higher anisotropy with multiple compression premoulding.

2.6 Composite properties

2.6.1 Mechanical properties

Generally the green strength of short fibre reinforced compounds is increased [61]. If the fibre content gets too high it can be that the two roll mill has to be stopped to be able to cut the sheet.

Tensile properties of short fibre reinforced elastomers like tensile strength and stress levels at small strains below 50% elongation [54, 59, 116] are increased in the fibre direction. This effect depends also on the fibre length and volume fraction similar to the findings for short fibre reinforced thermoplastics [117]. For low fibre contents a dilution effect lowers the modulus under the rubber compound without fibres, but from a certain fibre volume fraction level the tensile strength and modulus increase again. This critical fibre content for aramid fibre reinforced rubber was found to be varying with gum strength by Park and Kim [118]. It also depends on the polarity of the matrix and the cross-linking system used as shown for melamine fibres in EPDM by Rajeev et al. [24]. Kutty and Nando [51] reported that for short aramid fibre filled TPU the critical fibre content for a reinforcing effect was 10 phr. The same critical fibre content was found for short silk fibre reinforced NR/PE blends [59].

In transversal direction to preferred fibre orientation the Young's modulus is higher than the pure rubber compound but much lower than in longitudinal direction. When the fibre orientation is not uniaxial, any mechanical property measurement is the result of an averaged fibre orientation distribution. Strain at break values of the short fibre filled rubber compounds are decreased, because the fibres act as crack initiators in the rubber matrix.

Foldi [119] reported the compressive behaviour of short p-aramid fibre reinforced NR. A cylindrical test specimen was compressed to 60% of its original height. The fibres had to be perpendicular to the compressive force to be an effective reinforcement. The fibres in the mid-section of the sample were extended more than the fibres located at the endplates of the specimen.

In general tear strength of short fibre reinforced elastomers is increased for low fibre contents. It is also higher when the fibres are oriented transverse to the crack propagation direction. For high fibre volume fractions tear strength goes down because the strain amplification between closely packed fibres can promote tearing parallel to the fibre direction. Murthy [28] reported no change of tear strength by adding glass fibres to NR, while Cataldo [29] stated an reduction of tear strength of at least 20% by adding carbon fibres. Kumar and Thomas [58] found an increase of tear strength in both longitudinal and transverse direction with higher sisal fibre concentrations or residual fibre lengths in SBR.

Hardness increases generally with incorporation of fibres. Cataldo [29] could confirm this for carbon fibres in NR. Also Kumar et al. [120] showed that increasing the fibre content of sisal fibres in SBR increased the hardness and lowered the resilience. Ahmad and coworkers [48] mixed 3 mm short aramid fibres in a NR/HDPE blend. Maleic anhydride grafted polyethylene (PE-g-MA) was used as a compatibiliser to improve the fibre-matrix adhesion. They reported that aramid fibres behaved as an effective reinforcement especially in the system with PE-g-MA and hence increased the rigidity of the composites. However, further increase of fibre content showed that the hardness tended towards lower values. Gunnarsson and Nyman [47] showed that hardness was lower when the fibres were oriented in the direction of the indentation compared to the perpendicular orientation.

2.6.2 Dynamic mechanical properties

The addition of short fibres generally decreases and broadens the peak for tan δ and shift it with increasing fibre volume fraction to higher values. Figure 2.11 shows this trends for short aramid fibre filled TPU [52]. Ashida et al. [121] studied the dynamic moduli, E' and

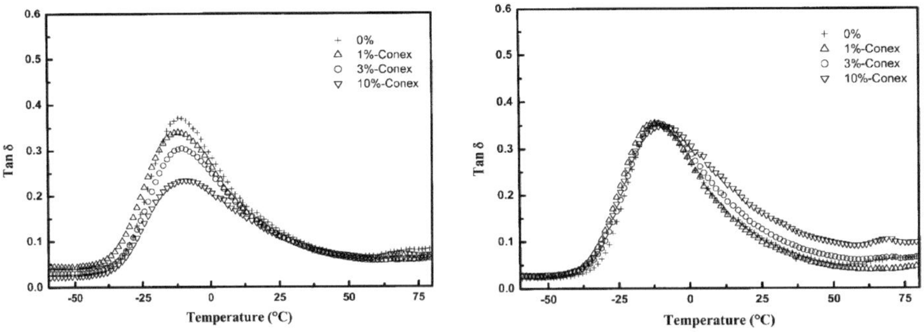

Figure 2.11: Tan δ versus temperature for m-aramid (Conex) in TPU composites as a function fibre content (left), normalised tan δ versus the temperature for Conex-TPU composites as a function of the fibre content [52] (right)

E'', and tan δ as a function of temperature for 6 mm nylon short fibres dipped with RFL and

PET short fibres dipped with isocyanate in CR composites. They investigated the influence of the fibre orientation and -content. Further they investigated [122] the influence of different dipped PET fibres in CR, EPDM, and urethane (UR) composites with unidirectional short fibres as a function of temperature. The decrease of tan δ in perpendicular to the fibre orientation was described by the equation to quantify the fibre matrix adhesion.

$$\frac{\tan \delta_{maxc}}{\tan \delta_{maxm}} = 1 - \beta\phi_f \qquad (2.36)$$

where $\tan \delta_{maxc}$ is the maximum $\tan \delta$ value for the composite for fibres oriented perpendicular to direction of measurement, $\tan \delta_{maxm}$ is the maximum $\tan \delta$ value for the matrix and ϕ_f is volume fraction of fibre. According to Ashida and co-workers a higher value of β indicates better adhesion between fibre and matrix.

Guo and Ashida [63] investigated the viscoelastic properties of short PET fibre filled TPE including styrene-isoprene-styrene copolymers (SIS) and styrene-butadiene-styrene copolymers (SBS). It was found that the level of storage modulus was rising by increasing the fibre content or length.

Ibarra [23] reported the dynamic properties of PET-fibre filled EPDM while varying the dynamic stress amplitude, fibre content and surface treatment of the fibre. Figure 2.12 shows storage modulus variation, in longitudinal measurements, E'_L, plotted against strain amplitude, in terms of double strain amplitude (DSA). All samples showed identical behaviour. A linear response region of the modulus, where it is independent from strain amplitude, was followed by a zone for which the modulus progressively diminished. By increasing the fibre content the linear response zone was being narrowed while the storage modulus increased. This phenomenon is well known for particle filled samples as Payne effect [21].

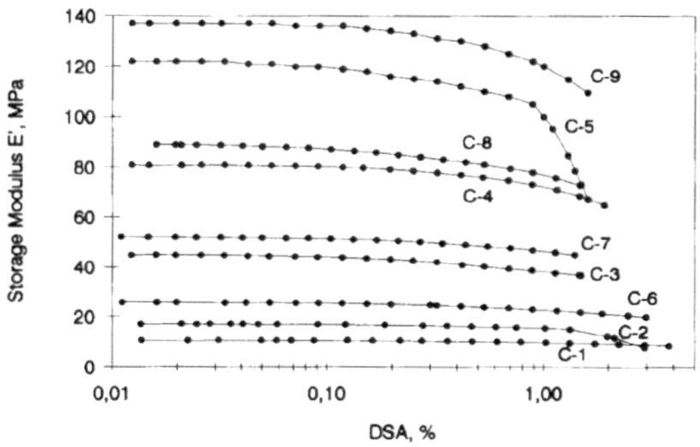

Figure 2.12: Storage modulus variation with strain amplitude. Measurements taken longitudinally to preferential fibre orientation in the matrix [23]

Another dynamic measurement is the so called heat build-up test. Its often measured by the Goodrich flexometer test and the specimen is exposed to an oscillating load with either constant peak stress amplitude or strain amplitude at high frequency. Then the temperature and compression set after testing are measured. Cataldo [29] stated that adding 1 phr of carbon fibres to NR reduced heat build-up and compression set of the resulting vulcanisates. Gunnarsson and Nyman [47] reported that an effect of orientation on heat build-up could not be measured. Contrary to that Rijpkema [46] stated the heat build-up was reduced about more than 80% for fibres oriented perpendicular to the direction of applied constant stress.

2.6.3 Abrasion, friction and wear properties

Friction and abrasion are interrelated, as abrasion is a process of wearing away the material surface by friction. Rubber abrasion is characterised by the formation of periodic surface pattern named abrasion pattern. It is well known that the abrasive wear of rubber is drastically reduced if the filler carbon black is used. Short fibres increase this effect. Foldi [119] showed the influence of fibre content and orientation of p-aramid pulp in a EPDM/CR blend. Gunnarsson and Nyman [47] reported that the results of DIN abrasion tester for short fibre reinforced NR/SBR blends were more depending on the currently used abrasive paper than on the fibre orientation. However, Rijpkema [46] could show that if various short fibre types were oriented perpendicular to the abraded surface, the abrasion was reduced significantly compared to the fibre free reference sample. The explanation is clear, when we consider that the fibres in this orientation have to be broken or pulled out. Contrary to that the fibres are easily torn out, when they are oriented parallel to the tribological surface.

The reducing of coefficient of friction by using aramid fibres in CR was reported by Uchiyama et al. [123]. They used samples with 10 and 20 vol% of short m-aramid fibres orientated perpendicular to the surface of a disc under dry and wet sliding conditions (Figure 2.13). They found that contact pressure reduces the surface roughness of the fibre reinforced compound. The coefficient of friction was dramatically decreased for both, wet and dry condition, depending on the contact pressure and speed. However, the desirable orientation

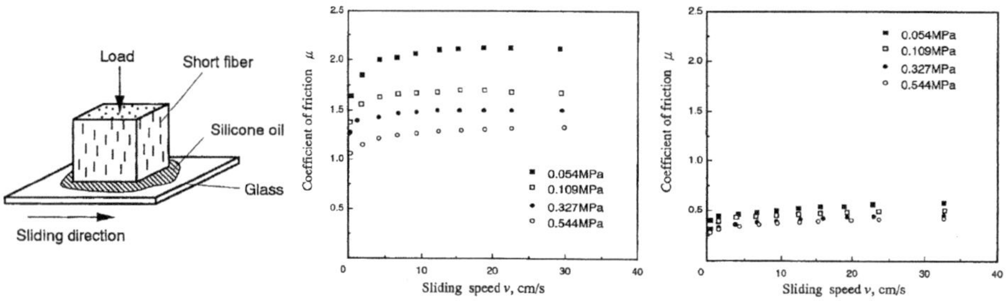

Figure 2.13: Configuration of the sliding pair (left) and variations in the dry friction coefficients with sliding speed under various contact pressures (middle) of unfilled matrix rubber and composite filled with 20 vol% m-aramid fibre (right) [123]

of the fibres perpendicular to the tribological surface is not easy achievable by standard rubber processing steps. Normally the compounds oriented on two roll mill are rotated 90° and stacked together to be moulded for the end application. The wear resistance of aramid fibres with different orientation in chloroprene rubber was also researched by Wada and coworkers [37, 124, 125]. Kashani [45] reported no significant abrasion loss difference for adding 1 phr of Technora to NR/BR compounds.

2.7 Characterisation methods

In this section a brief overview of the possible characterisation methods for short fibre reinforced elastomers is given. Morphology characterisation usually involves visual observation of a composite surface by light or electron microscopy. Due to the lack of transparency of most used elastomer compounds, characterisation methods include commonly also indirect measurements like mechanical testing in assumed fibre orientations and swelling methods [30, 31, 58]. But these methods interfere also with fibre-matrix interaction and are not feasible to separate the influences of morphology, like orientation and length from fibre-matrix interaction.

To detect the degree of fibre orientation usually surfaces of fractured tensile or tear specimen are investigated by scanning electron microscopy (SEM) in assumed main fibre orientation and perpendicular to it. Several authors used this method [24, 32, 34, 38, 59].

SEM is also commonly used to quantify the fibre-matrix interaction. Fibre treatments for enhanced adhesion and residuals of the matrix on the fibres are detectable and help qualitatively determine the fibre-matrix bonding. Young and co-workers [126] applied SEM to characterise the adhesion of surface treated p-aramid fibres to an epoxy resin. Similar studies have been carried out by Ahmad et al. [48, 49] and for TPU by Vajrasthira and coworkers [52]. Kumar et al. [41] used SEM to characterise morphology of short sisal fibre reinforced SBR extruded on an Instron capillary rheometer at different shear rates. SEM was also applied to determine the length of melamine short fibres after several processing steps in EPDM [24].

One of the newer microscopic techniques is atomic force microscopy (AFM). Rajeev and co-workers [24] studied the behaviour of melamine fibres with several dry bonding systems in EPDM compounds by AFM. The images confirmed the role of dry bonding systems in improving the fibre-matrix adhesion of the aged vulcanisate. De Lange et al. [90] investigated the effect of oily finish components on the interface strength of aramid-epoxy composites by AFM.

Raman spectroscopy is also applied to find out the micromechanical behaviour of aramid fibres composites under strain. Coffey et al. [95] investigated the interfacial micromechanics of p-aramid fibres in an engineering thermoplastic elastomer (polyether amide block copolymer) by determining the distribution of interfacial shear stress along fibres in single fibre model composites using Raman spectroscopy. Also Young et al. [126] used the shift of Raman spectrum of aramid fibre under strain for analysing the deformation micromechanics during single pull out tests.

Optical microscopy is often applied for fibre length distribution determination of processed fibres after dissolving the matrix [41, 51]. For detecting the fibre morphology this method is often not suitable, because of carbon black used in the formulations and time consumption through sample preparation. Coffey and co-workers [95] used an optical microscope to map visually fibre deformation and detected so called "apparent breaks".

An approach for determining fibre orientation in carbon black filled systems is the use of x-ray micrographs. Wada et al. [37] determined the orientation and dispersion of short aramid fibres in CR composites by x-ray photographs. The degree of short fibre orientation was characterised by a value, which calculated by two dimensional Fourier transformation of the soft X-ray photograph of the composite. X-ray was applied to 1 mm vulcanised thick sheets.

Due to the progress in non-destructive x-ray imaging techniques the achievable resolution increased during the years. Shen et al. [127] used it to determine internal structure in a phenolic foam reinforced with short glass fibres. The technique was used to measure the fibre length distribution (FLD) and fibre orientation distribution (FOD). Therefore they exploited the large difference in density between the two components. So the individual fibres were clearly visible on the 2D and 3D images generated. For quantitative analysis of morphology in

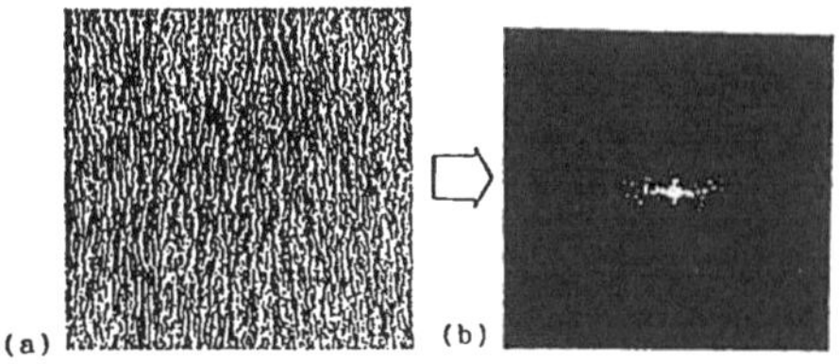

Figure 2.14: Soft x-ray photograph and its two dimensional Fourier transformation a) soft x-ray photograph; b) Two dimensional Fourier conversion [37]

short fibre elastomer composites, this computer tomography technique is more suitable as the conventional parallel dissection followed by image analysis of polished surfaces. The reason is that conventional approach is not feasible for curved high cut resistance fibres in elastomer composites. Sieker et al. [128] studied high resolution x-ray computer tomography performed with microfocus and nanofocus on polymeric materials. They showed that especially for rubber materials this technique can be applied. For example they presented the distributions images of zincoxide and carbon black in rubber samples. Also Alshuth and Robin [129] reported the possibility to apply this technique to elastomeric parts. They showed that studies of porosity in elastomers, analysis of the microstructure of fibre rubber composites and dispersion analysis are possible.

2.8 Applications

The field for applications of short fibre reinforced elastomers is increasing. Examples like tyres, belts, hoses, friction products and seals are well known and have reported by many authors [11, 107, 116, 119]. Van der Pol [7] and Clarke and Harris [116] reviewed several applications for short fibre reinforced elastomers and their feasibility for aramid fibres, which are dipped or in pulp form.

For tyres, especially the use of short fibres in a tread compound, van der Pol [7] summarised lower rolling resistance, better water drainage, and uniform wear as achievable properties. According to him, aramid short fibres give fewer mixing problems than high levels of high surface area carbon blacks in the bead area of a tyre. Other advantages are offered by the high anisotropy and the increased dynamic modulus, which lead to lower heat build-up. Life extension of the bead compound and preservation of the adhesion between bead wire and bead compound are also known. For off-the-road tyres the higher tear- and chunking resistance can be a big advantage. Pulp and dipped chopped fibres are suitable for this application. Clarke and Harris [116] proposed a fibre orientation of 45° as optimum for tyre treads, because it would give a high modulus and low hysteresis during travel, while offering low modulus and high hysteresis during skid.

According to Foldi [119] short fibre reinforcement in the cushion stock of a transmission belt has to be oriented perpendicular to the cord (and parallel to the neutral plane) to be effective. It prevents that individual cords cutting into the cushion stock, increases the side to side bending stiffness and also increase the wear resistance along the contact area with the sheave. Van der Pol [7] stated that the increased hardness of the rubber in fibre direction gives the transmission belt a lower friction coefficient and thus a reduced noise level when in service. Also a lower heat build-up during the cyclic compression and an increase in transmission capability are achievable.

Foldi [119] reported that conveyor belts should be reinforced perpendicular to the travel direction, because to prevent separation of the continuous reinforcing cables and of side-to-side sagging under payload. A side to side orientation of short fibres is preferable to reduce extension of the matrix across the width of the belt. Also the kind of pay load will determine what kind of reinforcement is needed and where is has to be located. For aramid fibres a fibre content of 3 to 5 phr was suggested.

According to Clarke and Harris [116] hoses are the most common applications for short fibre reinforced elastomers, because if the fibres are oriented in circumferential direction, the hose can withstand high bursting pressures and maintain at the same time its axial flexibility. Van der Pol [7] stated that small p-aramid pulp fibre contents (2–10 phr) showed better control of the volumetric expansion and improvement of the coupling retention. By using an expanding mandrel die the expansion of the rubber compound in the extruder head can force the fibres to change to the desired orientation. Goettler [115] processed straight and curved hoses of EPDM reinforced by treated cellulose fibre. He studied the influence of the die geometry on the fibre orientation in detail. Finally, Trelleborg [130] presented a curved hose filled with aramid pulp for automotive application (Figure 2.15). P-aramid pulp

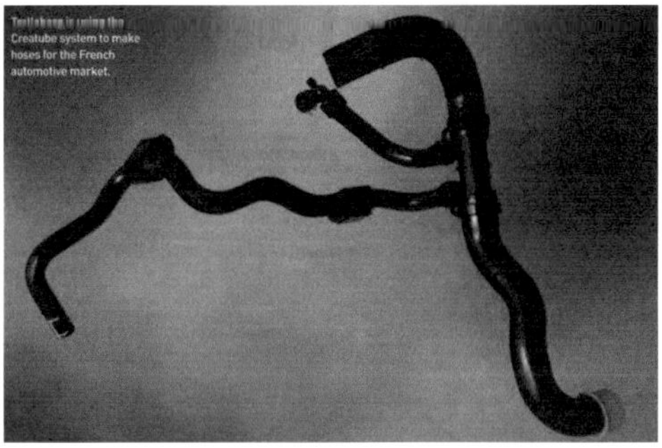

Figure 2.15: Trelleborg is using the Creatube system to make hoses for the French automotive market [130]

has also been used for a long time in the production of gaskets. Also for rubber seals and oil well packings the use of aramid pulp or dipped chopped fibres can bring advantages, like better mechanical properties at elevated temperatures, less creep, higher abrasion resistance and less swelling by solvents. The reduction of swelling is hereby highest in the direction of fibre orientation. Therefore a circumferential direction would be most effective.

Aramid in pulp form is used to substitute asbestos in friction products for automotive and train brakes and clutch facings, because of the pulmonary effect of asbestos. The favourable orientation for this application is a random one.

2.9 Patents

Most of the patents found were dealing with short fibre reinforced tyres. Brown and Scriver [131] of Goodyear filed a patent on rubber containing 0,2–20 phr aramid pulp reinforcement. Suitable applications were named as tyres, especially in carcass plies and reinforcing belts between treads and carcass in radial tyres. Nanni and Bruncacci of Pirelli [132] introduced an aramid pulp reinforced elastomeric layer between the belt structure and the tread band and the method of making the tyre. Causa [133] of Goodyear introduced elastomers contaning a mixture of partially oriented fibres and fibrillated pulp and their application in a tyre. Also patents with other fibre types or short fibres in general were claimed. Bergman and Arnold [134] of Michelin registered a patent on elastomer composites with glass microfibres. These elastomer compositions were suggested to be used for forming products that include a tyre tread or a support member within the sidewall of a tyre. The glass microfibres should be oriented in a substantially circumferential orientation or at a 45° orientation in a plane formed between the tread thickness and circumferential orientations or in a sidewall thickness direction. Görter et al. [135] of Semperit claimed a patent on a tyre comprising at least one layer formed of an elastomeric material with two layers of different oriented fibres. Boustany and Coran [136] claimed a patent on compounding discontinuous cellulose reinforced elastomers with the help of bonding agents. Another patent of them [137] was dealing with a pneumatic tyre reinforced by unregenerated cellulose short fibres. Moers [138] made a patent on a vehicle tyre which had a radial carcass with a reinforced crown and the tyre beads each comprise at least one beadwire surmounted by a filler strip. This was made of a rubbery mixture reinforced with short fibres, which were at least in part, orientated in the radial direction of the tyre. Finally, Guichard et al. [139] patented a reinforced silicone elastomer having polyamide fibres as reinforcement.

Chapter 3

Experimental

In this chapter the materials, devices and methods used for compounding, processing and characterisation are introduced and described in detail.

3.1 Materials

3.1.1 Elastomers

The two different base rubbers were ethylene-propylene-diene rubber (EPDM) and natural rubber (NR). Due to the limitation of the used characterisation methods to transparent compounds, beside each industrial carbon black filled elastomer compound also a transparent model system was investigated.

On the one hand EPDM formulations with peroxide curing system were used, containing a hose formulation as well as the model system, filled with curing system and different contents of fibres. On the other hand NR truck tread formulation with sulphur curing system and a model compound with peroxide curing system was chosen to guarantee the transparency of the NR model compound needed for optical characterisation. Table 3.1 shows the compound formulations used in this study. As usual in rubber technology the components are given in parts per hundred parts of rubber (phr).

The EPDM type used was Keltan 8340A from DSM Elastomers Europe B.V. with a Mooney viscosity of 80 MU $(ML1+4)$ 125°C, an ethylene content of 55%, an propylene content of 39,5% and ethylidene norbornene content of 5,5% [140]. The EPDM had an density of 0,86 g/cm^3.

NR used for the tread compound was a Standard Malaysian Rubber (SMR CV60) type with a Mooney viscosity of 60 MU $(ML1+4)$ 100°C. For the NR model system a Sri Lanka Rubber (Pale Crepe) type was used to obtain a high transparency. Pale crepe is a class of NR for which the impurities have been mostly removed [19]. The current type had Mooney viscosity of 99 MU $(ML1+4)$ 100°C.

The reinforcing fillers carbon black N220 and N550 were obtained from Evonik (formerly Degussa), Germany. Also plasticizers were used in the recipe. The oil type added to EPDM compounds was paraffinic oil type Sunpar 2280 and in case of NR it was a naphthenic oil type Nytex 840. Several other antioxidants, antiozonants and vulcanisation additives were added to the compounds. ZnO with Red Seal quality was coming from Union Minière, Belgium, and stearic acid was a technical quality grade.

Polyethyleneglycol (PEG) was obtained from Merck, Darmstadt, Germany. Sulphur and

Tertiary-Butyl-BenzothiazolSulfenamide (TBBS) were provided by Rhein Chemie, Germany. Trimethylol-propane-trimethacrylate (TRIM) and Perkadox 14/40 peroxide were obtained from AkzoNobel, Deventer, the Netherlands. Polymerized 1,2-dihydro-2,2,4-trimethylquinoline (TMQ) was supplied from Flexsys, Belgium. Dicumylperoxide (DCP) was obtained from MERCK, Hohenbrunn, Germany.

Component	EPDM model	EPDM hose	Component	NR model	NR tread
Keltan 8340A	100	100	NR	100	100
N-550		105	N-220		55
Stearic acid		1	Stearic acid		2
Sunpar 2280		60	ZnO		5
PEG		2,5	Nytex 840		8
TMQ		1,25	TMQ		1,5
Perkadox 14/40	1,5	7,5	6PPD		2
TRIM 50	0,6	4	Wax		2
			TBBS		1,5
			PVI		0,1
			Sulphur		1,5
			DCP	2	
Twaron	0 - 5	0 - 5	Twaron	0 - 1	0 - 5
Technora	0 - 5	0 - 5	Technora	0 - 1	0 - 5

Table 3.1: Used compound recipes, ingredients in parts per hundred parts of rubber (phr)

The fibre wt% and vol% for the compounds were calculated according to appendix 6.1 and are shown in table 3.2. Due to the different compositions of the formulations, the resulting vol% differed even the phr were constant. The vol% of the model systems was always higher compared to the industrial system, because the fibre part was diluted by the other components of those compounds.

		Model		Industrial			
		EPDM	NR	EPDM hose		NR tread	
phr	wt%	vol%	vol%	wt%	vol%	wt%	vol%
1	0,97	0,58	0,68	0,35	0,27	0,56	0,46
2	1,92	1,16	1,34				
3	2,85	1,73	2,00	1,06	0,81	1,65	1,37
4	3,77	2,30	2,65				
5	4,67	2,86	3,29	1,75	1,35	2,72	2,26

Table 3.2: Calculation of wt% and vol%

3.1.2 Aramid fibres

Two different fibre types namely Twaron (PPTA) and Technora (PP/ODPTA) were used in this study. Twaron was dipped by three different fibre treatments with different fibre lengths. One treatment was an oily standard finish, another a standard treatment plus epoxy layer and the last a resorcinol formaldehyde latex (RFL) dip, was consisting of the epoxy layer and the RFL layer. Technora just had standard and RFL dip. In the following the dips will be abbreviated with "stan"', "epoxy" and "RFL". On macroscale the colour differences between the different treatments can be seen for Twaron in figure 3.1. Standard and epoxy bundles obtained yellow colour caused from the aramid itself. Single filaments were sticking out of the bundles indicating a rather low cohesive strength. Contrary to that RFL bundles showed brownish colour due the colour of the thicker dipping layer. The cohesive strength of the bundles was higher since the filaments were glued together by the dip. Both fibre types were used as supplied by Teijin Aramid B.V. in plastic bags.

Figure 3.1: Images of Twaron fibre with stan, epoxy and RFL dip

3.2 Compounding and Processing

Preparation of short fibre reinforced elastomers is similar to that of conventional rubber compounds. The typical processing ways feasible for rubber and thermoplastic elastomers like internal mixer, two roll mill, extruder also can be applied. The main difference is that additional aspects like fibre dispersion, fibre breakage and fibre orientation become more important. In this study fibres were mixed into the compounds in internal mixer or on two roll mill. The incorporation on internal mixer is hereby better reproducible in terms of fibre dispersion and fibre length, because for two roll mill mixing the operator influence is higher. In case of model systems the basic processing influences were investigated on internal mixer. Curatives always were added on two roll mill to avoid undesired curing during long time mixing in the internal mixer. Figure 3.2 shows the scheme for the incorporation of fibres. For the industrial compounds the masterbatches were mixed on 150 l internal mixer without curatives. The fibres and curatives were then added on laboratory scale two roll mill.

3.2.1 Internal mixer

Two different sized internal mixers were used for fibre dispersion studies. On laboratory scale a 78 ml mixer (Haake 610p) was used for most of the experiments. The larger internal mixer

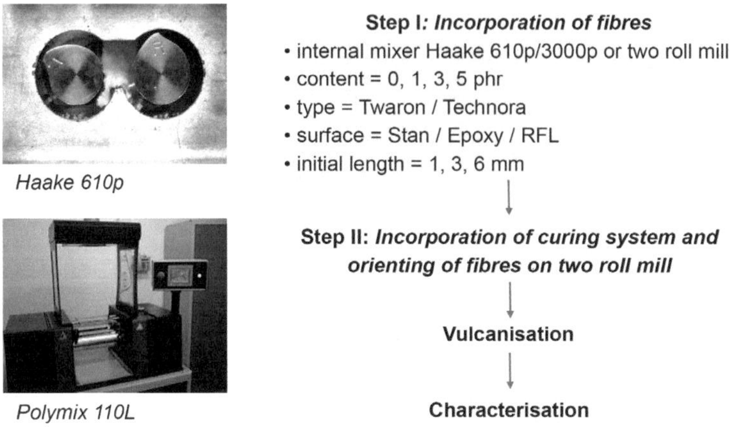

Figure 3.2: Scheme for incorporation of fibres for model systems

with 378 ml (Haake 3000p) was used to do upscaling experiments. The mixers were equipped with Banbury rotors and are shown in figure 3.3.

For the Haake 610p internal mixer (fill factor of 0,7) the standard parameters were set to at 90°C for EPDM (110°C for NR) with a rotor speed of 40 rpm with the mixing sequence: 50% matrix for 2 min, after that fibres and 50% matrix in for t_{mix}, dump.

Experiments varying the fill factor, mixing time and rotor speeds, as well as experiments using a glass window to visualise the mixing process were carried out to study the influence of the processing conditions.

Figure 3.3: Haake 610p (left), Haake 3000p (right)

3.2.2 Two roll mill

Fibres or curatives were added on a two roll mill type Polymix 110L, from SERVITEC Maschinenservice GmbH, Wustermark, Germany. Compounds were, if not other specified, processed at room temperature with roller speeds set to 17 and 14 rpm, constant friction ratio of 1,2 and a nip gap of 0,5 mm. The matrix was masticated for 2 min and then the fibres were incorporated. After incorporation of fibres and curatives the fibres were oriented by friction ratio of 1,0 and same nip gap. The obtained sheets were then prepared for vulcanisation by compression moulding.

3.2.3 Compression moulding

Vulcanisation was always carried out by compression moulding with a hot press from Fortijne Grotnes BV, AC Vlaardingen, the Netherlands at a pressure of 10 MPa and at a temperature of 170°C for EPDM compounds and NR model formulation or 140°C for sulphur cured NR tread compound. For standard characterisation methods like tensile testing and dynamic mechanical analysis 2 mm thick sheets were pressed with a standard mould with geometry of 114 mm x 94 mm x 2 mm corresponding to optimum cure time t_{90} + 2 min of the compounds.

For heat build-up measurements cylindrical samples with a diameter of 18 mm and a height of 25 mm were pressed by piling up slices punched out of the milled sheets. The temperature and pressure were kept 1 h for EPDM or 1,5 h for NR samples to secure full vulcanisation. The fibre orientation was in the direction of the circular surface.

In case of specimen for friction testing all cylindrical samples with a diameter of 60 mm and height of 10 mm were cured for 2 h with corresponding temperature and pressure in a special constructed mould with 6 cavities. Figure 3.4 shows this mould. To achieve different preferential fibre orientations in the disc specimen the sheets obtained from two roll mill were cut into stripes in direction or perpendicular to main fibre orientation and stapled accordingly into the mould. Also random orientation was introduced by taking the compounds directly from the internal mixer after processing it together with the curatives. During compression moulding there was melt flow on one side, which could not be avoided and had an influence on the fibre orientation. Therefore the application of the non-affected disc side was recommend for friction testing.

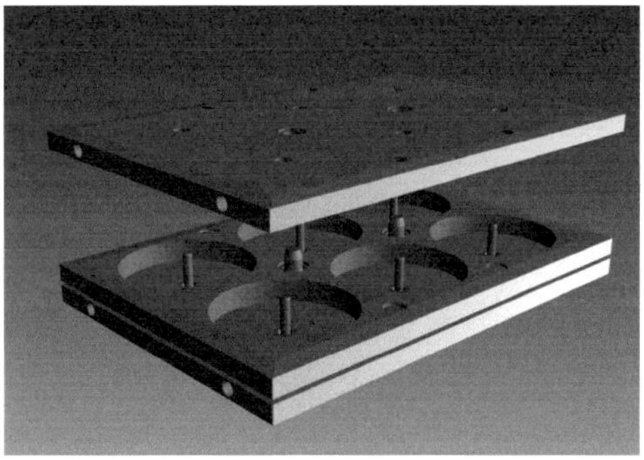

Figure 3.4: Mould with 6 cavitities to prepare specimen used in tribological tests

3.2.4 Design of experiments

The dependence of the bulk properties of short fibre reinforced elastomers on several parameters are shown schematically in figure 3.5. For the transparent model systems the influence of fibre, mixing and processing parameters were investigated. Mixing and processing parameters like used device, applied time, total strain and specific energy input were studied as well as fibre parameters like content, initial length and surface treatment. These influence factors led to a compound morphology, which was characterised in the following by specific values for fibre dispersion, length and orientation distribution. The quasistatic, dynamic and tribological properties were then determined with respect to the found morphology.

The abbreviations ETE and OTR mark hereby the main objectives for the other two parties involved in the SUSTFIBER project and stand for the departments of Elastomer Technology and Engineering and Surface Technology and Tribology, both University of Twente, the Netherlands.

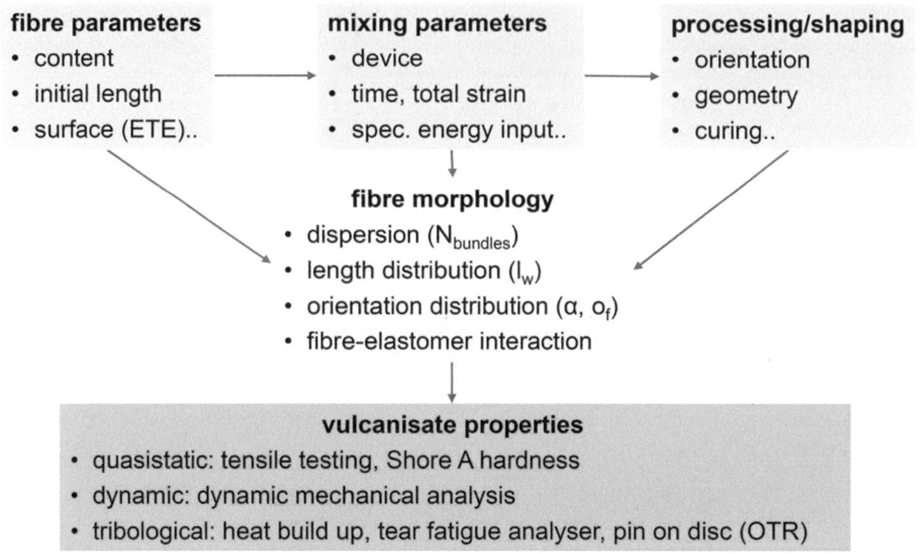

Figure 3.5: Scheme for influencing factors on bulk properties of short fibre reinforced elastomers

3.3 Characterisation methods

3.3.1 Characterisation of fibre dispersion

Experiments regarding fibre dispersion were just carried out for transparent model systems. Trials to detect the fibre dispersion for carbon black filled compounds by an ultraviolet light lamp failed due the high absorption of the light through the carbon black.

After incorporation of the fibres for specific time on internal mixer or two roll mill the material was two roll milled to sheets with a surface of 800 cm^2 and a thickness of 0,5 mm. These sheets were inspected manually to quantify the macroscopic dispersion. All objects with a diameter larger than 3 mm were counted and defined as the number of undispersed fibre bundles $N_{bundles}$. Trials with recording the degree dispersion by photography and automated detection by software were hampered by the low contrast and surface roughness of the sheets. Photos of the vulcanised specimen were made to visualise the dispersion. Figure 3.6 shows examples for undispersed fibre bundles in two roll milled sheets (left) and vulcanised samples (right).

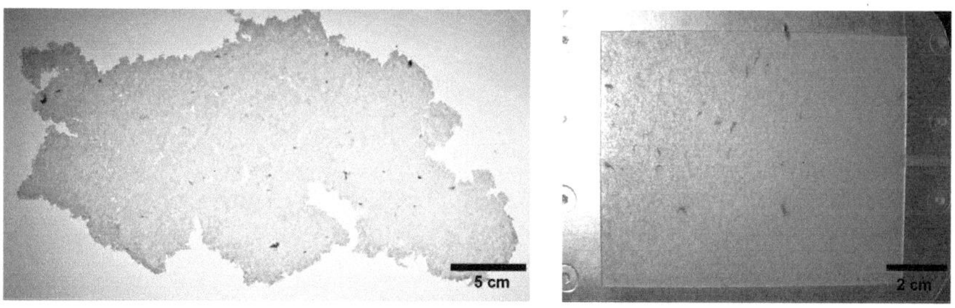

Figure 3.6: Examples for undispersed fibre bundles in two roll milled sheets (left) and vulcanised samples (right)

3.3.2 Characterisation of fibre length distribution

The initial fibre length before processing was determined by measuring 100 random fibre bundles with a digital calliper.

To characterise the fibre length distribution (FLD) of the fibres in the model compounds, transparent films with a thickness of 15 μm were compression moulded from the unvulcanised material. In the next step fluorescence microscopy was executed. Fluorescence microscopy is widely used in biology to visualise structures of micro-organisms, but often the samples have to be labelled with fluorescent stains to enable the technique. Due to the aromatic rings of their chemical structure the aramid fibres were fluorescing themselves when illuminated with UV-light. The high contrast of the images together with their high resolution made it possible to differentiate single filaments and even separated fibrils.

A microscope in reflection mode (Axio Imager A1m from Carl Zeiss Jena GmbH, Germany) was used to record images of these films with a magnification of 5x, which were later compiled with software (MethaMorph SPOT Insight) to 18,4 mm x 14,4 mm images. The fibres were originally emitting blue light. To enhance the contrast a threshold was applied to the original images and the fibres were marked green.

The size of these images was limited by the datasize of the compiled images and the area of the films. According to Fu et al. [141] it is also possible to correct the fibre length distribution by the fibres intersecting the recorded images. They demonstrated that for short glass fibres in thermoplastics.

However, for this study fibre lengths distribution was determined by counting manually the length of minimum 200 fibres fully within the image. Objects touching the edge and objects with a diameter deviating from the normal thickness were left out. Automated detection of fibre length with software, as it is standard for glass fibres after burning away the matrix and disperse the fibres in aqueous solution [142], failed due to curvature and crossing points of the aramid fibres.

An advantage of the applied procedure over the conventional method is that the fibre length was not further changed, because the steps of burning away the matrix and dispersing the fibres again could be omitted.

For comparison of the fibre length distribution obtained, characteristic values were calculated as follows. The number average of fibre length l_n is calculated after:

$$l_n = \frac{\sum N_i \times l_i}{\sum N_i} \tag{3.1}$$

The weighted average of fibre length l_w is defined as:

$$l_w = \frac{\sum N_i \times l_i^2}{\sum N_i \times l_i} \tag{3.2}$$

Where l_i is the product of i multiplied with a finite increment of fibre length Δl, while i ranges from 1 to n and N_i is the number of the fibres with that fibre length l_i.

To summarise figure 3.7 shows a compiled image (left) and the fibre length distribution with determinded characteristic values (right).

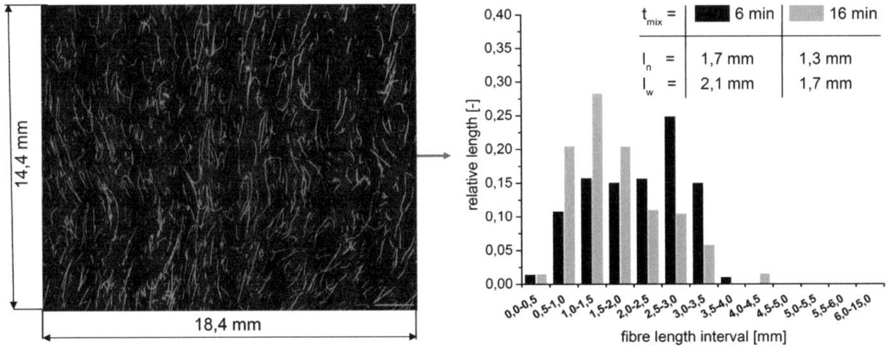

Figure 3.7: Procedure for determination of fibre length distribution

3.3.2.1 EPDM and NR industrial compounds

Since the carbon black was hampering the fibre detection, methods used for characterisation of carbon black dispersion were investigated to be applied for detecting short fibres in carbon black.

There are two main methods for evaluating the carbon black dispersion. The microtomed section method and the torn surface method [143]. Thereby for the microtomed section method frozen rubber samples have to be sectioned and then the area of undispersed agglomerates is characterised by light microscope leading to a percentage of dispersed carbon black. For the torn surface method a compound in which ingredients are perfectly dispersed is torn apart. The light is then reflected from the torn surface giving a smooth texture. If there is improperly dispersed carbon black it is indicated by irregularities.

Since whole fibres should be detected, it had been focussed on the microtome method. According to Medalia [144] the sections had to transmit enough light. If the carbon black contents are very high the sections thickness has to be less than $2\,\mu m$. If we consider that aramid fibres used were 12 μm in diameter, there was no way to fulfil this condition. Trials with slices having $15\,\mu m$ thickness were not successful as it can be seen in figure 3.8. The fibres were just visible partly and no reproducible FLD could be executed.

In another attempt it was tried to remove the matrix by two ways. The matrix was either burned away at 400°C for 30 min in a microwave ashing furnace MAS 7000 from CEM or dissolved by toluene using a Soxhlet extractor. The cycle time was hereby 10 min. The weight loss was measured every 4 hours. After 2 days the matrix was extracted and no more

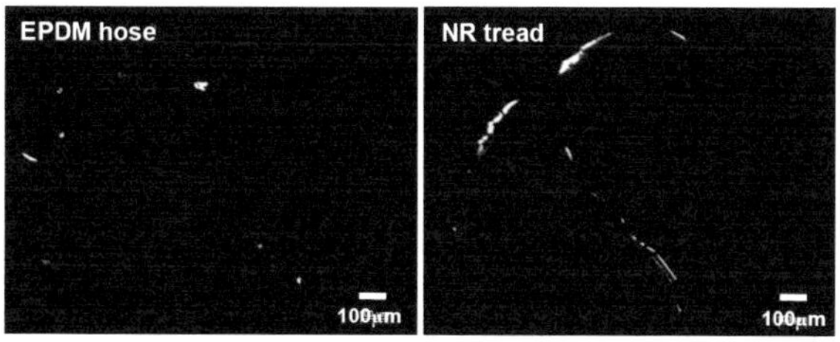

Figure 3.8: 15 μm thick slices of EPDM hose (left) and NR tread (right) with 1 phr standard dipped Twaron fibres

weight loss could be monitored. The influence of carbon black still hindered the capture of adequate images. Several observable fibres parts were disconnected by overlying carbon black particles. Due these problems only a few fibres could be observed, so that the FLD could not been statistically confirmed. Trials to obtain the fibre length distribution by using a x-ray technique failed since the volume of the scannable sample was too small for sufficient fibre length characterisation.

3.3.3 Characterisation of fibre diameter distribution

For determining fibre diameter distribution the same films (see figure 3.7) as for the fibre length distribution were investigated. On the obtained images the diameters of 30 objects were detected. Due to the scattering of the fluorescence light the objects appeared thicker than they originally were. Therefore a correction coefficient was introduced to normalise the filaments with an average of 12 μm.

3.3.4 Characterisation of fibre-fibre interaction

To determine fibre-fibre interaction films compression moulded for the fibre length distribution were studied. For the obtained images a threshold was applied to binarise them. The overall surface area of the detected objects (including also the fibrils) was measured then automatically by Scandium Software. This fibre surface area was then divided by the overall surface area of the investigated image.

3.3.5 Characterisation of fibre orientation

3.3.5.1 Confocal laser scanning microscopy of model systems

Confocal laser scanning microscopy (CLSM) is a well-established technique to visualise the 3D morphology of short fibre composites, especially for glass fibre reinforced thermosets and thermoplastics [10, 145]. Similar to fluorescence microscopy the aramid fibres fluoresce strongly when excited by the argon-ion laser, but contrary to it the higher z-resolution and reduced blur makes confocal pictures more precise and clearer.

To determine the orientation of the model compounds, five 2 mm thick S2 tensile bars were punched out of a vulcanised sheet. The middle section (3.15 x 6.3 mm^2) of these tensile bars was then scanned with CLSM (TCS SP5, Leica Microsystems, Wetzlar, Germany) using magnification of 10x and were later compiled to 3D high resolution overview stacks by software. The penetration depth of the laser was limited to $400\,\mu m$ depending on the light loss or distortion of the investigated matrix. A projection of the 3D fibre microstructure in plane $100\,\mu m$ below the surface was processed by binarisation. To get the autocorrelation this image was then processed by the FD Math algorithm from ImageJ software with a macro similar to works of Pegel et al. for carbon nano tubes in polycarbonate. Details can be found in the corresponding paper [146]. To obtain the main angle of fibre orientation and the orientation factor the binarised image was investigated with a covariance or two point probability function. This function could be interpreted as the area fraction of the intersection of the binarised image with the image shifted by the vector r (figure 3.9 a,b). Applying this procedure for all discrete vectors r in the range of the image size led to an intensity map as shown in figure 3.9 c. In the example the anisotropic character led to a

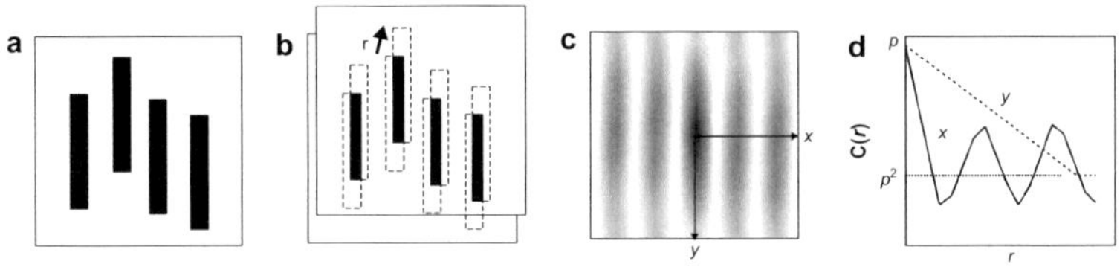

Figure 3.9: Example for a covariance analysis of a given random closed set [146]

strongly elliptical intensity distribution. The regularity of the given distribution was reflected by the periodicity of intensity in x-direction (figure 3.9 c and d). An orientation factor o_f was then calculated indicating the ratio of the correlation functions at a distance r:

$$o_f = 1 - \frac{x(r)}{y(r)} \tag{3.3}$$

Thus the highest orientation would result in a o_f of 1, while for random orientations it would be zero.

Finally, the main orientation angle and its deviation α from the direction of tension in the tensile test bars were determined. This angle α and the orientation factor o_f were then correlated to the Young's modulus obtained from tensile testing. To summarise the procedure a sketch is drawn in figure 3.11.

3.3.5.2 Scanning electron microscopy

The fibre orientation and interaction with elastomer was investigated with different magnified scanning electron microscopy (SEM) images of the surface of fractured tensile bars. SEM was executed with the help of Ultra 55 plus (Carl Zeiss SMT, Germany) scanning electron microscope with an acceleration voltage of 4 kV. Before examination samples were sputtered with 2 nm thick layer of platinum.

3.3.5.3 High resolution x-ray computed tomography

As reported light microscopy was not feasible to detect the fibre morphology in carbon black filled compounds. Therefore high resolution x-ray computed tomography was executed. This technique allows to scan a specimen with submicron resolution and visualise the morphology in 3D even for highly filled elastomer samples [128, 129]. First the sample was rotated and series of two dimensional x-ray images were recorded. After that an image inversion and filtering were carried out and a back projection was generated. Finally, these back projections were compiled to an virtual 3D volumetric data model, which shows possible pores, cracks or filler distributions. From the gained virtual reconstruction it was possible to derive 2D slices in any direction.

Samples were sent to Phoenix x-ray (GE Sensing and Inspection Technologies, Wunstorf, Germany). An EPDM hose sample filled with 5 phr Twaron standard fibres and the geometry of 1 x 1 x 1 mm^3 was scanned by high resolution computer tomograph (nanotom) with tungsten target. The stepsize was 1,5 μm and therefore the scanning time was 3 h. With this high resolution x-ray technique it was possible to distinguish the fibres and their orientation and length even in the carbon black filled EPDM compounds. Figure 3.10 shows the resulting virtual cut and the reconstruction. Further studies with this technique could not be obtained due to the high costs of external measurements and the long scanning times. However, this technique seems to be capable tool for investigation of morphology of carbon black filled rubber compounds.

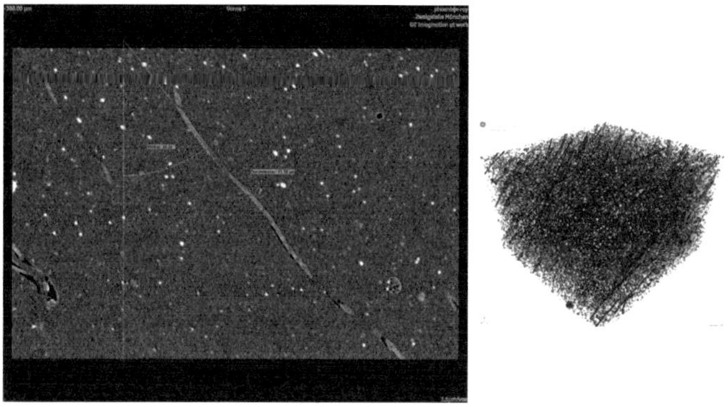

Figure 3.10: High resolution computed tomography for EPDM hose compound with 5 phr Twaron standard fibres: virtual cut (left) and cubic reconstruction (right)

3.3.6 Characterisation of fibre-elastomer interaction

To determine the fibre-elastomer interaction high magnification SEM images (detail described in section 3.3.5.2) on the surface of fractured tensile bars were investigated.

3.3.7 Curing characteristics and Mooney viscosity

To characterise the Mooney viscosity and curing behaviour of the studied compounds, specimens were tested after DIN 53523 [106] and DIN 53529 [147].

Curing values, like minimum (S'min) and maximum torque (S'max), curing times t_{10} and t_{90} were determined with the help of rubber process analyser (Scarabaeus SIS V50) from Scarabaeus Mess- und Produktionstechnik GmbH, Langgöns, Germany. The specimen were tested according to DIN 53523 with an amplitude and frequency of 0,2 ° and 0,83 Hz, respectively. EPDM and NR model compounds were tested at a temperature of 170°C, while NR tread compounds were tested at 140°C. From the obtained values $\Delta S'$ and t_{90}-t_{10} were calculated.

Mooney viscosity is defined as the shearing torque resisting rotation of a rotor surrounded by rubber in a cylindrical cavity and was tested according to standard DIN 53523. Deviating from the standard the temperature was switched for model compounds to 90°C, the initial temperature of the internal mixing chamber. The results were written as ML(1+4) 90, where ML denotes Mooney viscosity measured by using a large rotor, 1 is for 1 min of preheating, 4 is for 4 min of shearing, and 90 is for 90°C, the temperature of the chamber. It should be noted that the recipe of the chosen NR model compound without fibres was close to standard test compound for control of rubber process analyser according to DIN 53529 part 2.

3.3.8 Characterisation of surface energies

Contact angle data for the fibres was supplied by Teijin Aramid B.V. using a FTA188 Video Tensiometer measuring three contact angles per fibre-dip combination and liquid. The two liquids were water and diiodomethane. The measurement was executed for dynamic condition with 60 frames per second by applying the liquids on yarn bobbins. The surface free energy was then calculated according to Wu [148].

Contact angle measurement of the unvulcanised matrix rubbers was carried out using a Dataphysic OM Micro device measuring ten contact angles after Laplace-Young per material and liquid. The six liquids were: formamide, water, ethanol, ethylene glycol, ethylene glycol-water mix (ratio 20 : 80), ethanol-water mix (ratio 10 : 90). The measurement was done in static mode. The surface free energy was calculated according to Owens-Wendt-Rabel-Kaelble (OWRK) [149, 150].

3.3.9 Tensile testing and shore hardness A determination

The tensile properties were determined according to DIN 53504 [151] using five S2 type tensile bars, a cross head speed of 200 mm/min using a 1 kN load cell installed at tensile testing machine from Zwick GmbH, Ulm, Germany. Secant modulus between 1 and 2% strain defined as Young's modulus, stress and strain in percentage elongation were recorded during the testing. Later the measured Young's modulus was compared with that calculated by Halpin-Tsai model and lamina theory using l_w and α and the corresponding values from table 6.1. To summarise the procedure a sketch is drawn in figure 3.11. The sketch shows

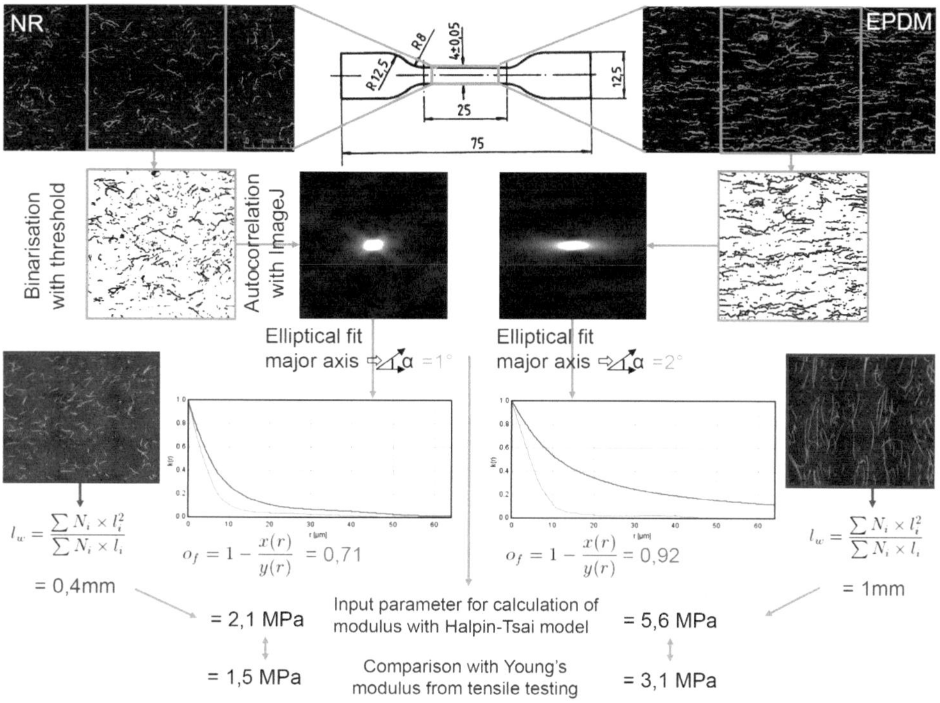

Figure 3.11: Schematically sketch of the procedure to compare calculated with the measured Young's modulus

that for the two given examples a low deviation angle α was determined. However, this value can be misleading since α is an average over all fibres in the investigated area and can be also low in case of a random orientation. Therefore the orientation factor o_f may help to judge how relevant α is. Together with l_w a modulus can be calculated according to equation 2.16 which is then compared with the measured one.

Hardness values were measured according to DIN 53505 [152], using a Shore A Durometer from Heinrich BAREISS GmbH, Oberdischingen, Germany. Five indentations were made for each material by piling up three tensile bars.

3.3.10 Dynamic mechanical analysis

Dynamic mechanical analysis (DMA) is an useful tool to measure the damping ability of short fibre reinforced elastomers. Hereby vibration is induced with a certain frequency, amplitude and over a temperature range. The response is monitored and the storage modulus E', loss modulus E" were calculated. The loss factor tan δ, which is the quotient of E" divided through E' had a peak related to the glass transition temperature. Williams, Landel and Ferry [153] found that there is a law for determining frequency-temperature equivalence, the so called WLF equation. Because viscoelastic properties of polymers cannot be measured for high frequencies due to limitations of the testing equipment, it is possible to measure these properties at low temperatures which are equivalent to those measured at high frequencies.

Therefore the frequency-temperature dependent viscoelastic mechanical properties of the vulcanisate including storage E', loss modulus E'' and loss factor $\tan\delta$ were measured in a temperature range from -80°C to 80°C at a heating rate of $2\,K/min$ and constant frequency. For the measurements a dynamic mechanical analyser, EPLEXOR 2000N (Gabo Qualimeter Testanlagen GmbH, Ahlden, Germany) using a 2 mm thick rectangular rubber vulcanisate block having an area of 35 x $10\,mm^2$ was used. The measurements were done in tensile mode with strain amplitude and frequency being set to 0,5% and 10 Hz, respectively. Static prestrain was set to 1%, corresponding to a low elongation where the fibres were still stretched but not pulled out. Figure 3.12 shows the specimen used for DMA (left) and heat build measurements (right) and the EPLEXOR 2000N (middle).

Figure 3.12: DMA specimen (left), EPLEXOR 2000N (middle), heat build-up specimen (right)

To investigate the amplitude dependence of the viscoelastic mechanical properties of short fibre reinforced elastomer vulcanisates the same sample geometry was tested under amplitude variation from 0,1 to 100% strain with 100% static prestrain.

3.3.11 Heat build-up

DIN 53533 specifies flexometer test for the determination of the temperature rise and resistance to fatigue of vulcanised rubbers. Many rubber products, such as tyres and belts, are tested by subjecting them to an oscillating load with either a constant peak stress amplitude or strain amplitude. In order to obtain good correlation between accelerated tests and in-service exposure of these products measurement of heat build-up and failure behaviour should be carried out. Samples (figure 3.12 right) were prepared according to DIN 53533-3 [154] and cured for 1 h (EPDM) and 1,5 h (NR). The fibres in the test specimen were hereby oriented perpendicular to the direction of applied stress. One sample was tested for each material. Before every test the geometry was measured. Tests were executed in compression mode and started at room temperature for the condition of constant dynamic strain (4,45 mm) or constant dynamic stress (0,8 MPa) respectively. During the 25 min of measurement the temperature was online measured on the top of the sample. After each test the temperature inside the sample and compression set after 1 h was recorded.

3.3.12 Tear fatigue analyser

The service lifetime of rubber articles can by characterised by special testing procedures. Many methods as the DeMattia tester or the DIN abrasion device obtain no correlation between the predicted behaviour and the real crack growth resistance of the studied compounds [47]. Therefore the influence of different orientations for short fibre reinforcement on fatigue crack-growth behaviour was studied by a adequate testing device, a tear fatigue analyser (TFA). Fatigue tests were carried out with notched specimens under cyclic tension. The used tear fatigue analyser (Coesfeld, Germany) was modified in order to simultaneously test tensile and pure shear mode specimen. However, in the current experiments only pure shear mode specimen were used since Radek [155] could show that this mode is better reproducible. Figure 3.13 shows the arrangement of the tensile and pure shear mode specimen in the TFA device and the geometry of the pure shear mode specimen used. EPDM hose

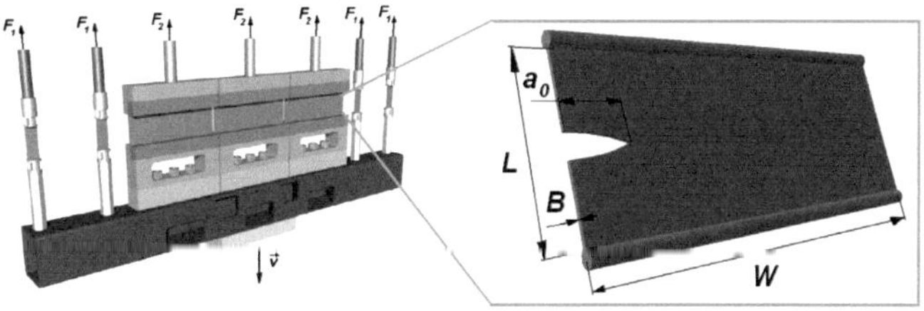

Figure 3.13: Arrangement of the tensile and pure shear mode specimen in TFA device (left), pure shear mode specimen (right) [156]

compounds without and with 5 phr Twaron standard fibres oriented parallel and perpendicular to the crack propagation direction were tested. The pure shear test specimen had the following geometrical dimensions: height L = 15 mm, length W = 120 mm, thickness B = 1,5 mm, initial cut a_0 = 30 mm. The test conditions for the specimen were set to a preforce of 1 N and a frequency of 10 Hz. The amplitudes were ranging from 4,9 til 19,68 %. Tearing Energy for the pure shear mode specimen was calculated according to the works of Rivelin and Thomas [157]:

$$T = -\frac{dW_E}{B \cdot da}\bigg|_{L=const.} = L \cdot W_E \tag{3.4}$$

where $dW_{Elastic}$ is the amount of which the complete elastic energy of the specimen is changed when the crack propagates by da at fixed grip conditions, while W_E is the elastic strain energy density undisturbed from the crack front and the right edge of the specimen [157]. There is a linear relationship of double logarithmic plots of crack propagation rate and the tearing energy. In a certain regime a linear relation between crack propagation rate of a cycle can described according to Paris and Erdogan [158] with:

$$R = \frac{da}{dn} = f(T) = C \cdot T^m \tag{3.5}$$

where a is the crack length, n is the number of cycles and T is the maximum tearing energy in the current cycle, while C and m are often interpreted as material constants.

Chapter 4

Results and discussion

In the following chapter the results are reported and discussed regarding the raw materials, their compounding and dispersion behaviour, the fibre aspect ratio, the fibre-elastomer interaction, curing behaviour, fibre orientation and the physical properties of the resulting composites.

4.1 Characterisation of raw materials

4.1.1 Initial fibres

The initial length of the fibres as delivered was determined by measuring 100 fibre bundles with a digital calliper. The results can be seen in table 4.1. For Twaron the initial fibre length distribution showed a similar average value for all measured fibre types slightly above the expected value of 1, 3 or 6 mm respectively. The fibres with standard finish and 3 mm desired length showed higher scattering containing to a small extent fibre bundles with lengths higher than 10 mm. This was attributed to problems in the cutting process. Because aramid fibres have a high cutting resistance, they are often applied in protective clothing. Another effect is that the knifes which cut the fibres are getting blunt and the desired length of the fibre bundles is exceeded. Contrary to the increased deviation for Twaron, the standard deviation for Technora fibres was significantly lower, because a different cutting process was used in this case [159].

Type		Twaron							Technora		
Surface		stan		epoxy		RFL			stan	RFL	
l_n	mm	3,29	6,36	3,36	6,57	1,39	3,26	6,17	2,96	3,22	6,16
l_w	mm	3,79	6,44	3,47	6,63	1,51	3,31	6,18	2,97	3,23	6,17
s		1,28	0,71	0,60	0,66	0,41	0,38	0,27	0,15	0,16	0,27
Min	mm	2,43	2,93	0,96	3,42	0,76	2,62	5,64	2,61	2,88	5,41
Max	mm	10,46	7,20	5,84	8,77	3,18	4,79	6,91	3,56	3,76	7,40

Table 4.1: Initial fibre length measured for 100 random fibre bundles

To determine the fibre diameter and the surface roughness scanning electron microscopy (SEM) images of the initial fibre surfaces were recorded and are shown in figure 4.1. These images were then further analysed by line scans of their grey value to evaluate surface roughness and diameter.

The initial fibre diameter was varying between 10 and 15 μm for all investigated fibre types and surface treatments. The surface roughness increased hereby for Twaron from the standard finish (a) with smooth surface over epoxy (b) to RFL dip (c) with uneven distribution of the dip. As already observable for the images of the fibre bundles in figure 3.1 some of the filaments were glued together by the RFL dip, which can be seen in figure 4.1 (f). For Technora the surface roughness was in general higher compared to Twaron. No significant differences between standard (d) and RFL dip (e) were noticed in this case.

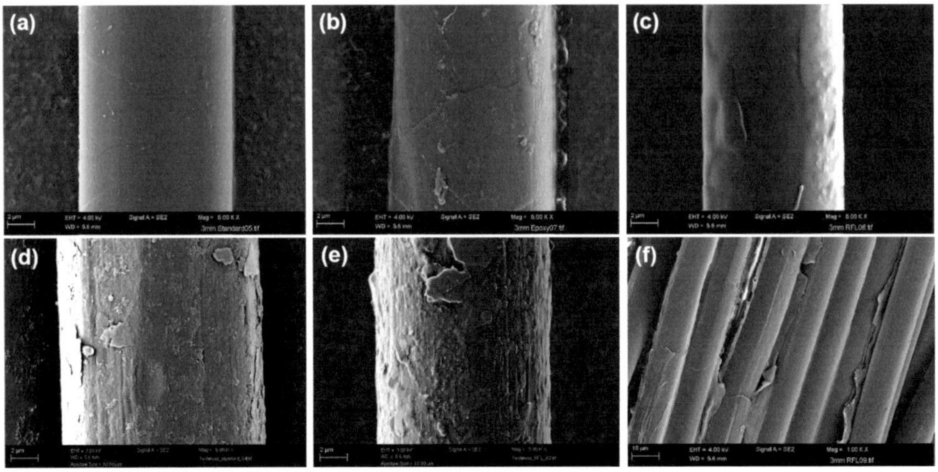

Figure 4.1: SEM images of initial Twaron fibres with standard (a), epoxy (b) and RFL dip (c,f) and Technora fibres with standard (d) and RFL dip (e)

4.1.2 Surface energies of fibres and matrices

Contact angles measured on yarns and calculated surface energies γ with disperse and polar parts are shown in table 4.2. For Twaron the polarity decreased from standard over epoxy to RFL dipped yarns. Technora fibres with standard finish already had a non-polar fibre surface. Technora dipped with RFL could not be measured since just short fibres and no yarns were available. Contrary to the findings of the fibre yarns, the unvulcanised model elastomers did

Material	Angle (°)		γ (mN/m)		
	H_2O	CH_2I_2	Total	Disp.	Polar
Twaron (raw)	87,0	24,2	51	46	5
stan	32,7	26,7	75	45	30
epoxy	68,5	42,2	53	39	14
RFL	112,3	55,4	31	32	0
Technora (raw)	107,2	33,6	41	43	0
stan	99,0	47,6	38	36	2

Table 4.2: Contact angles measured and calculated surface energies γ for fibre yarns

not show significant differences (see table 4.3). Both showed non-polar behaviour.

	Angle (°)						γ (mN/m)		
	CH_3NO	$C_2H_6O_2$ $+H_2O$	C_2H_6O $+H_2O$	$C_2H_6O_2$	C_2H_6O	H_2O	Total	Disp.	Polar
EPDM	87,1	97,2	91,8	90,4	29,0	117,6	22,0	21,9	0,1
NR	95,9	94,6	102,1	96,9	26,6	114,5	21,6	21,6	0,0

Table 4.3: Contact angles measured and calculated surface energies γ for unvulcanised elastomers

From both tables only the trends can be compared to predict the dispersion behaviour. For Twaron the non-polar fibre dip RFL should disperse better in both model systems compared to the more polar epoxy and standard finish. In case of Technora the standard finish showed already non-polar behaviour which should lead to no dispersion problems.

4.1.3 Mooney viscosities

The measured and supplied Mooney viscosities are shown in table 4.4. Mooney viscosity was tested with $ML(1+4)$ according to standard DIN 53523. Different to the standard the temperature was switched for model system to 90°C, the corresponding initial temperature of the internal mixing chamber. For the two EPDM types K2340A and K8340A also Mooney viscosities for a temperature of 125°C were displayed according to product data sheets [140, 160]. The standard type EPDM Keltan K8340A had the highest viscosity followed by the NR Pale crepe, the 50-50 mix of the two EPDM types and the low viscous EPDM type Keltan K2340A. For the industrial systems Mooney viscosity was measured for a temperature of 100°C according to standard. The NR tread compound had a lower viscosity of 47 MU compared to the EPDM hose compound with 60 MU.

EPDM	90°C	125°C		100°C
K8340A	131	80	NR Pale Crepe	99
K2340A	50	25	EPDM hose	60
50-50 mix	84		NR tread	47

Table 4.4: Mooney viscosities for model and industrial systems

4.2 Compounding and fibre dispersion

Dispersion was studied for the transparent compounds while varying different processing and material parameters. Therefore experiments were carried out to study the influence of mixing time, fill factor and rotor speed. For the material parameters fibre contents, types and dips were investigated. To follow the mixing process experiments were carried out on Haake 610p internal mixer with a glass window. A camera was installed to record two pictures every second. The resulting images can be seen in figure 4.2. As material EPDM model system filled with 1 phr Twaron RFL fibres was chosen, since the contrast for the other dips was too low. On the first image the empty mixing chamber with rotors can be seen. Next picture shows the mixing chamber filled with 50% EPDM, followed by the image where the fibres and the rest of the matrix were introduced. For increasing mixing time, the compound is getting more homogeneous in colour. The counter-rotating tangential rotors distribute and disperse the fibres into the material at the middle area between them by every rotation. Of course this front view can just give a partly overview on the chaotic flows in the internal mixer, but this illustrates that the mixing process is clearly dependent on the mixing time.

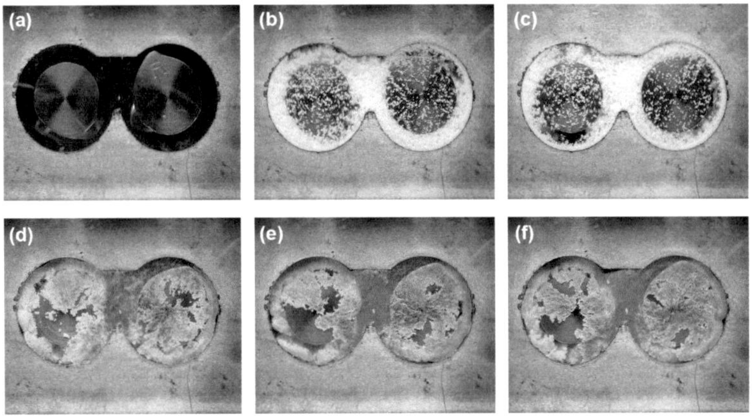

Figure 4.2: States of dispersion for K8340 + 1 phr Twaron RFL 3 mm in dependence of mixing time 0 (a), 1 (b), 2 (c), 3 (d), 4 (e) and 5 min (f)

First of all the influence of different processing parameters was investigated for 1 phr Twaron in EPDM standard type Keltan K8340 in internal mixer Haake 610p with free mixing volume of 55 ml, rotor speed of 40 rpm and a starting mixer wall temperature of 90°C. With a low degree of fill (DOF) of 0,5 and mixing time of 6 min a fibre bundle free compound could be created. By changing to a DOF of 0,9 and holding the mixing time constant the dispersion was insufficient. Therefore DOF was set to 0,7 for the following experiments. Similar value was reported as optimal DOF by several authors [107, 111]. The initial temperature was set to 90°C, since for higher temperatures the dispersion was insufficient. Also the rotor speed was varied in that sense that the total strain Γ, which is the product of the shear rate $\dot{\gamma}$ and mixing time t, was kept constant. Experiments were carried out by varying the rotor speeds with 20, 40 and 80 rpm and the corresponding fibre mixing times, t_{mix}, with 16, 8 and 4 minutes. No undispersed fibre bundles could be found for all these cases.

Dispersion of fillers consists of time dependent rupture and erosion processes. It is influenced by the matrix viscosity, by the cohesive strength and the wettability of the matrix polymer [19]. For example a possible infiltration of fibre bundle works better for low viscosity matrices with same polarity as the filler. Contrary the transfer of shear forces to overcome the filler cohesion works better for high viscosity matrices. To investigate the influence of matrix viscosity on the number of dispersed fibre bundles $N_{bundles}$ 1 phr of Twaron fibres with standard finish were incorporated into the standard EPDM type with 80 MU (ML 1+4 at 125°C), a low viscous EPDM type K2340 with 25 MU and a 50-50 mix of both. The detailed Mooney viscosities were reported in table 4.4. Figure 4.3 shows the torque and temperature behaviour in Haake 610p and the $N_{bundles}$ versus mixing time and specific energy input for different matrix viscosities and 1 phr Twaron fibre with standard finish. With increasing viscosity also the torque and temperature levels were rising. Therefore the specific energy input was increasing with higher viscosities since it is dependent on the integrated surface under the torque curve.

Contrary to the mentioned mechanism the influence of viscosity on the dispersion was not that significant as expected. Both, the high viscosity EPDM Keltan K8340A and the low viscosity EPDM Keltan K8340A had similar dispersion after same specific energy input. Same for the 50-50 mix of both. Generally the dispersion was improving with rising mixing time or specific energy input.

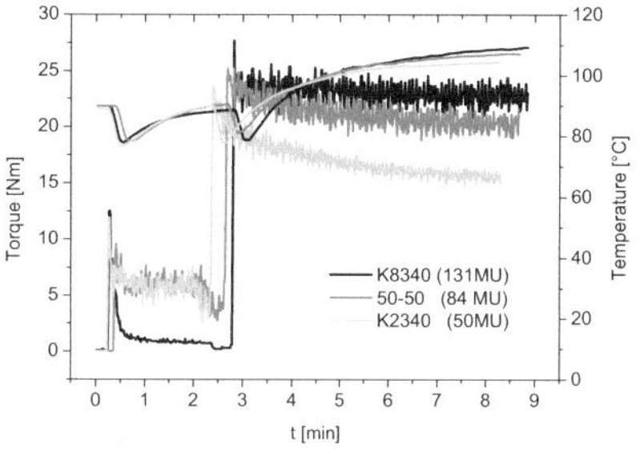

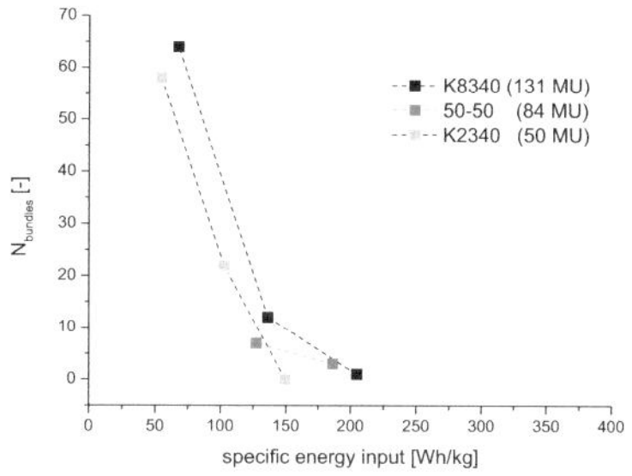

Figure 4.3: Torque and temperature in Haake 610p and $N_{bundles}$ versus mixing time for different matrix viscosities for 1 phr Twaron with standard finish and internal mixer with 0,7 degree of fill and 40 rpm rotor speed

Next figure 4.4 shows the recorded torque and compound temperature in Haake 610p and $N_{bundles}$ versus mixing time and specific energy input for EPDM model system filled with various Twaron fibre contents and 1 phr Technora with standard finish. With increasing Twaron fibre content also the torque and end temperature level went up. This is attributed to the higher friction between the fibres and the matrix [59]. Between the two fibre types with 1 phr fibre content no differences were observable. For higher Twaron fibre contents of 3 and 5 phr the specific energy input needed to reduce $N_{bundles}$ increased. Comparing 1 phr of Technora and Twaron it was clear, that Technora needed more input to disperse. Since the polarities of the fibres and the EPDM (tables 4.2 and 4.3) suggested the opposite behaviour another mechanism should be the reason. A possible explanation is a different fibre breakage behaviour, because longer fibres need more time to disperse [40, 61].

In a next step the fibres were mixed using the internal mixer Haake 3000p with a free volume including rotors, V_0, of 265 ml to investigate upscaling behaviour. Because of the

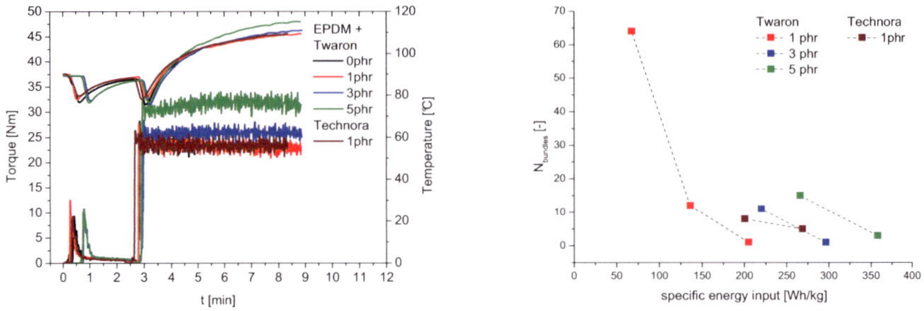

Figure 4.4: Torque and temperature in Haake 610p and $N_{bundles}$ versus mixing time for EPDM model system filled with various Twaron fibre contents and 1 phr Technora

different geometries different temperature and behaviour was obtained for the same standard parameters of mixing. Figure 4.5 shows the comparison of the two used internal mixers Haake 610p and Haake 3000p for 1 phr fibre content of both fibre types. Like expected the measured torque and wall temperature differed quite signifcant. While the average torque in Haake 610p was 25 Nm, in Haake 3000p it jumped to a value of 160 Nm. Also the temperature rise for the Haake 3000p was higher. This was caused by the lower surface to volume ratio compared to Haake 610p. With respect to the different fibre dips no difference, neither in torque nor in temperature curves, could be observed. To transfer the results of the small internal mixer to the bigger one, for the condition of constant dimensionless dispersion number N_{RS}, the corresponding optimum fibre mixing time t_{mix} was calculated according to equation 2.30 and the specific geometrical data of the used mixers displayed in table 4.5. Also shear rate $\dot{\gamma}$, total strain Γ and specific energy input e were calculated according to equations 2.32, 2.33, and 2.35 (page 24) for both mixers and are shown. For the calculated t_{mix} of 11 min similar dispersion results as for the Haake 610p could be achieved for the Haake 3000p. The maximum shear rate between the rotor tip and the mixer wall $\dot{\gamma}_{max}$ was more than doubled, while the total strain Γ was four times higher. The specific energy input e needed for sufficient fibre dispersion was 2,4 times higher for the bigger mixer (figure 4.5 right).

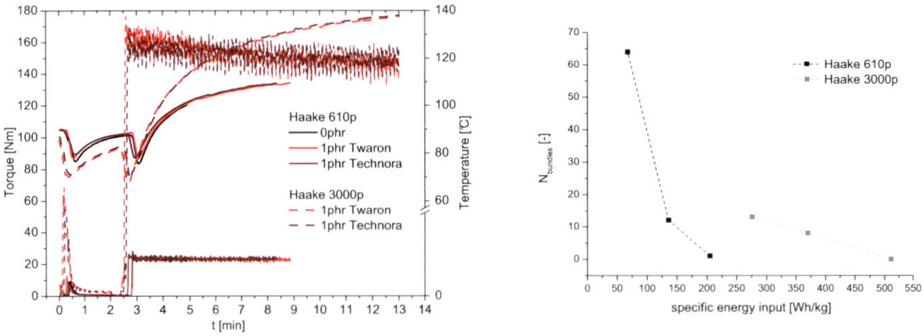

Figure 4.5: Torque and temperature in Haake 610p or Haake 3000p and $N_{bundles}$ versus mixing time for EPDM model system filled with 1 phr Twaron and Technora

Internal mixer type		610p	3000p
Volume with rotors V_0	ml	55	265
Degree of fill	%	0,7	0,7
Batch weight m	kg	0,0475	0,2288
Rotor speed ratio		1,50	1,50
Rotor tip clearance h	mm	1,7	1,4
Rotor radius R	mm	17,9	32,6
Rotor length L	mm	47,0	83,5
Rotor speed N	rpm	40	40
Fibre mixing time t_{mix}	min	6	11
Dispersion number N_{RS}		41	41
Shear rate $\dot{\gamma}_{max}$	1/s	44,1	97,5
Total strain Γ		15878	64376
Specific energy input e	Wh/kg	203	485

Table 4.5: Geometrical and calculated data of used internal mixer

The dispersion behaviour for 1 phr fibre with different surface modifications in EPDM and NR was investigated after that. It should be noted that due to dispersion problems the initial mixing temperature for NR Pale Crepe was increased to 110°C instead of 90°C like for EPDM. Figure 4.6 shows the comparison of torque and temperature in Haake 610p versus mixing time for EPDM and NR model system filled with 1 phr Twaron fibre in the upper middle section. Below $N_{bundles}$ is displayed versus specific energy input for each matrix. Although different elastomer types were used the temperature behaviour was similar with offset of 20°C, between the initial temperature of 90° and 110° for EPDM and NR, respectively. The temperature value was for both matrices dropped by 10 K when the fibres were introduced. In the end of the mixing process the temperature increased about 30 K. However, for the torque behaviour differences were found. EPDM torque value dropped to zero after the first 50% were introduced, while NR value fall down to 3 Nm. After adding the remaining 50% and the fibres EPDM showed a stable torque of 25 Nm, while NR dropped from a similar level under 20 Nm. This behaviour is most likely caused by the NR mastification process induced by the input of mixing energy and rising temperature.

The number of undispersed bundles $N_{bundles}$ versus specific energy input for Twaron and Technora with different surface treatments in EPDM and NR is displayed in the lower part of figure 4.6. As shown the shape of the temperature curve for both matrices was similar. Therefore the viscosity decrease between the addition of the fibres until the end of the mixing should be also similar. When the fibres were incorporated with 50% of the remaining matrix the resulting higher viscosity should help to erode the filaments from the fibre bundles. The temperature increase during mixing lowers the viscosity and therefore is pronouncing the infiltration of the remaining bundles leading to rupture.

In general all fibre types and treatments can be dispersed in EPDM at significant lower fibre mixing times or specific energy input compared to NR. This can be explained by the higher elasticity of the NR, which hampers the infiltration ability of the matrix into the fibre bundles. Because of that erosion seems to be the most favoured dispersion process in NR compared to EPDM, where rupture and erosion contribute to dispersion.

The RFL dipped Twaron fibres showed in both matrices incomplete dispersion in form of agglomerates containing several filaments. This was unexpected since from contact angle

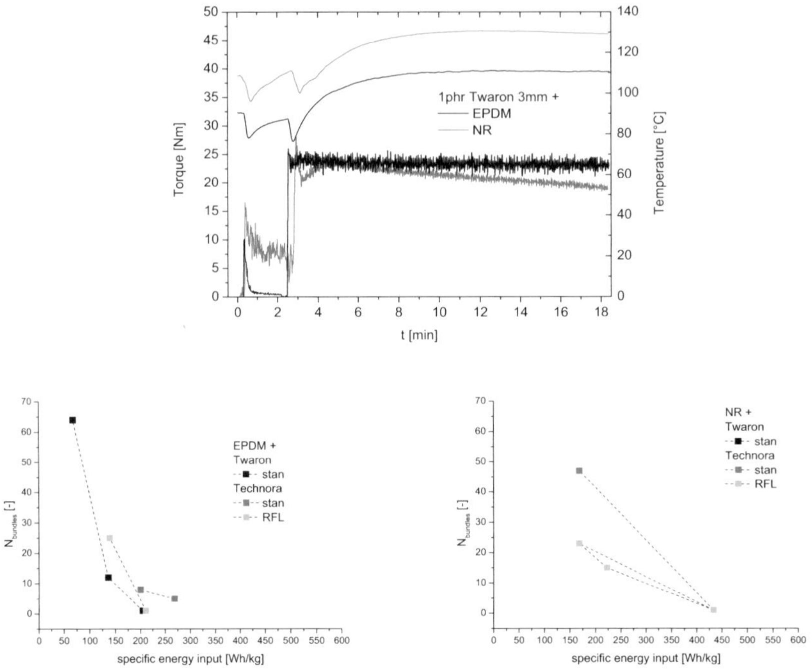

Figure 4.6: Comparison of torque and temperature in Haake 610p versus mixing time (upper middle) and $N_{bundles}$ for EPDM (bottom left) and NR model (bottom right) system filled with 1 phr fibre content

measurement (table 4.2) this finish was reported to be non-polar similar to the matrices. The reason for the insufficient dispersion was coming from the dip process itself. The uneven distribution of the dip on the filaments originated from the dip procedure. The applied stresses during mixing were too low to overcome the cohesive strength of the filaments with the highest dip coverage on the outside of the yarn. The agglomerates were even observable by eye and under microscope, which can be seen in figure 4.7 b and d, where the dispersion after 6 min for standard parameters is shown. However, for the high mixing times in NR the number of agglomerates was significantly reduced from hundreds to less than ten. In case of Technora fibres this behaviour was not observed. Here the RFL dipped fibres dispersed even better than the standard dipped fibres.

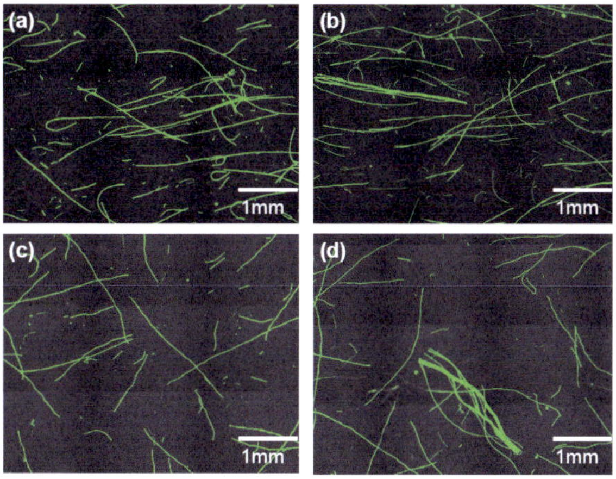

Figure 4.7: Detail images for dispersion after 6 min mixing: EPDM with Twaron Stan a) and RFL finish b); NR with Twaron Stan c) and RFL finish d)

Summary

Dispersion was investigated for the transparent compounds while varying different processing and material parameters. Experiments were carried out to study the influence of fill factor, specific energy input and mixing device. In case of material parameters different matrix viscosities, fibre contents, types and surface treatments were investigated.

The fibre dispersion was generally dependent on the specific energy input. Increasing the fibre content or length was causing a higher specific energy input or mixing time needed to get a sufficient state of dispersion. Upscaling the dispersion quality from different sized internal mixers was possible, but the needed specific energy input was enlarged. No significant influence of viscosity could be detected for EPDM. The used fibres dispersed earlier to filament level in EPDM compared to NR, because its elasticity was lower. For Twaron with RFL dip agglomerates or lumps containing multiple filaments were observed.

4.3 Fibre aspect ratio

4.3.1 Fibre length distribution

The fibre length distribution (FLD) after processing for the transparent model compounds was obtained by fluorescence microscopy as described in section 3.3.2. The dependence of fibre length indicators l_n, the number average of fibre length, l_w, the weighted average of fibre length on the number of counted fibres is displayed for an example in figure 4.8 on the left side. With growing number of counted fibres from 10 to 200 l_n went down from over 2 to 1,6 mm. The standard deviation for l_n, stayed constant with increasing fibre number. l_w followed the trend of l_n at a higher level starting at 2,4 mm going down to 2,1 mm. For 200 counted fibres an adequate compromise between statistical reliability and time consumption for counting the fibres manually could be achieved. Therefore for each investigated compound minimum 200 fibres were counted. The repeatability of different fibre contents of 3 mm standard dipped Twaron fibres mixed in internal mixer is observable together with 2 images for 1 and 5 phr in figure 4.8 on the right. The repeatability was sufficient with l_w values between 2,1 and 2,3 mm and standard deviation of 25 to 60 μm even for high fibre overlap in case of 5 phr.

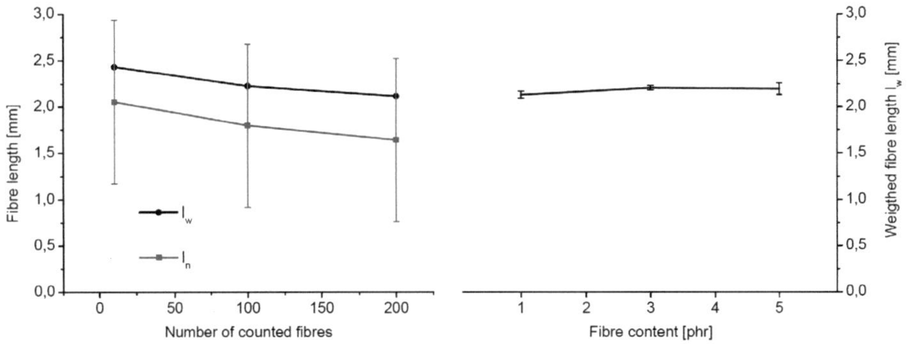

Figure 4.8: l_n and l_w in dependence of number of counted fibres (left), influence of Twaron fibre contents on repeatability of l_w values for 200 measured fibres after mixing in EPDM (right)

FLD was studied similar to fibre dispersion by varying different processing and material parameters. To investigate the influence of processing parameters, variations of mixing time, rotor speeds and mixing devices were studied. Also material parameter like fibre contents, types and dips were investigated. The important influence factors on l_w are observable in the following graphs.

First the influence of degree of fill (DOF) on l_w and dispersion was investigated for Haake 610p. Figure 4.9 shows the number of undispersed fibre bundles $N_{bundles}$ and l_w versus the DOF. By increasing DOF from 0,5 to 0,7 the l_w value and $N_{bundles}$ remained constant. For a DOF of 0,9 with less mixing space $N_{bundles}$ jumped up to 35, while l_w dropped to 1,9 mm. Therefore a DOF of 0,7 was chosen for the following experiments.

In the next step the influence of total strain on weighted fibre length l_w for the mixing on internal mixer with DOF of 0,7 and 90°C initial temperature was investigated for a matrix

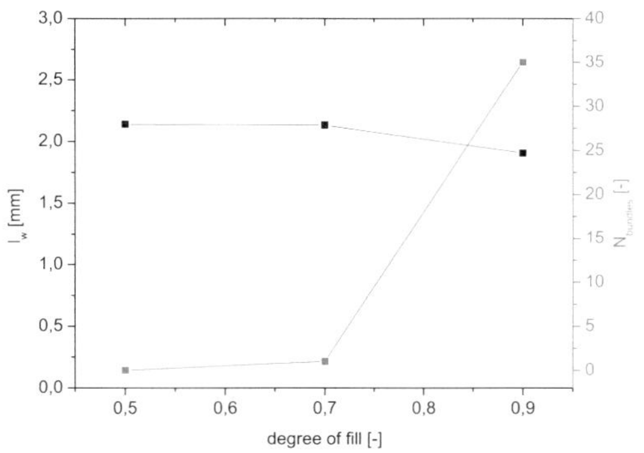

Figure 4.9: $N_{bundles}$ and l_w versus degree of fill for Twaron mixed in Haake 610p internal mixer

filled with 1 phr fibres with standard dip. For Twaron the highest residual fibre length l_w of 2,1 mm could be obtained for combination of a rotor speed of 40 rpm and mixing time of 6 min corresponding to a total strain of 15878. For increasing the total strain, the product of the shear rate $\dot{\gamma}$ and fibre mixing time t_{mix}, the l_w were decreasing to around 1,7 mm irrespective of using a high rotor speed with short mixing time or low rotor speed with long mixing time. Over a total strain of 20000 l_w stayed constant. In contrast to this the fibre breakage of Technora fibres was negligible. They kept relatively high residual fibre length of about 2,6 mm even for high applied total strain.

The comparison of $N_{bundles}$ and l_w versus total strain for Twaron and Technora fibres mixed in EPDM is displayed in figure 4.10. As total strain grows the dispersion is getting better and $N_{bundles}$ is going down. For Twaron this happens earlier than for Technora. One reason could be that longer fibres need more time to disperse [40, 61] and the increasing total strain decreases l_w for Twaron, while Technora is staying nearly constant. The distinct chemical structure of the two fibre types (figure 2.4, page 13) is the origin of the different breaking behaviour.

Next step was to investigate the influence of initial fibre lengths on the resulting fibre lengths after mixing. This is an interesting topic, since unnecessary fibre breakage can be avoided by finding the optimal initial length for a system. Figure 4.11 shows l_w after mixing for 6 min on internal mixer for different initial fibre lengths, types and dips. The dotted diagonal line represents the optimum line where no fibre breakage occurs and l_w after mixing is equal the initial l_w.

Twaron with initial lengths of 3 mm length could keep higher l_w values after mixing compared those with 6 mm initial length. As already mentioned the fibre breakage level for Technora was significantly lower compared to Twaron. Therefore l_w values of Technora came closest to the optimum line. Also Twaron with RFL dip and 1 mm length came close to the optimum. Generally all Twaron fibres treated with RFL obtained higher l_w values compared to the other dips. This can be attributed to the protective effect of the RFL dip against fibrillation. Even the RFL dip was a only a thin layer (figure 4.1 c) it might haved helped preventing crack initiation in the filament skin which is the starting point for fibrillation

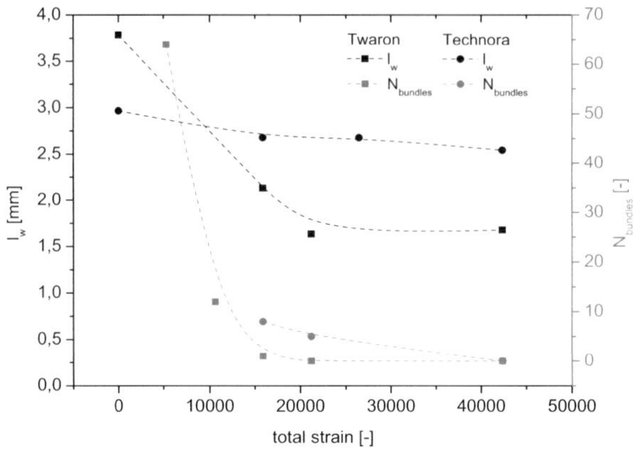

Figure 4.10: $N_{bundles}$ and l_w versus total strain for Twaron and Technora mixed in EPDM Haake 610p internal mixer

and fibre length reduction [75]. Also the found agglomerates (figure 4.7 b and d) played an important role, since they can withstand buckling during the mixing better.

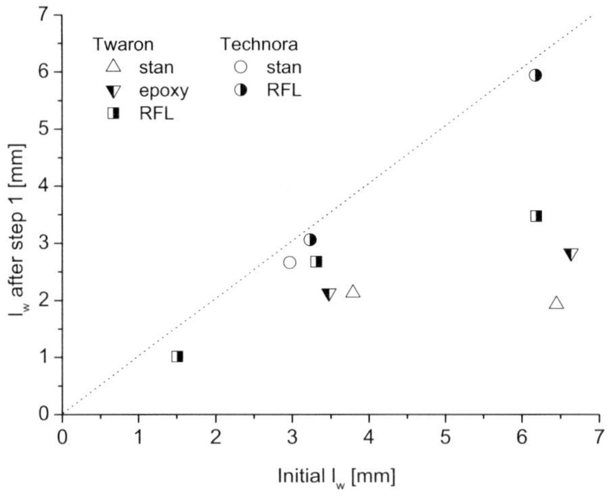

Figure 4.11: Influence of initial fibre lengths on l_w values after mixing step 1 of EPDM model system on internal mixer

To summarise the dispersion and fibre breakage behaviour of Twaron in internal mixer and different matrices, figure 4.12 shows $N_{bundles}$ and l_w versus total strain. As mentioned

$N_{bundles}$ decreased earlier in EPDM compared to NR, while the l_w dropped with increasing total strain and reached for EPDM a plateau of about 1,7 mm above a total strain of 20.000. For NR the l_w decreased to a level of 1,5 mm for highest applied total strain and compound with sufficient dispersion. As mentioned the different elasticities of the matrices caused the different times needed for sufficient dispersion.

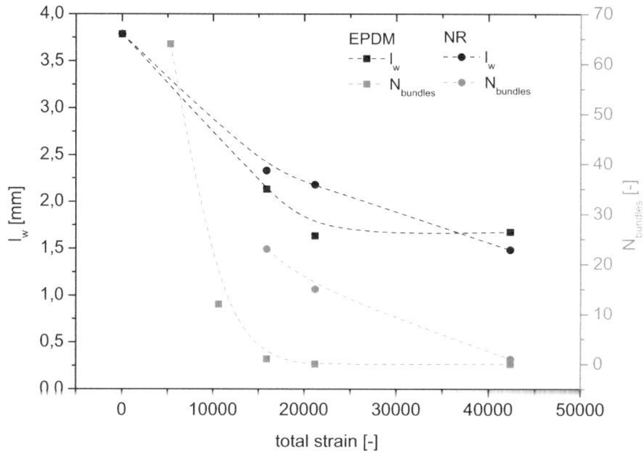

Figure 4.12: $N_{bundles}$ and l_w versus total strain for standard dipped Twaron with standard finish mixed in EPDM or NR on Haake 610p internal mixer

In the following the dispersion and mixing behaviour of Twaron fibres were investigated regarding the different mixing devices. Figure 4.13 shows $N_{bundles}$ and l_w versus the specific energy input for the different sized internal mixers Haake 610p or Haake 3000p.

To compare the dispersion and shortening behaviour of Twaron fibres between two roll mill and internal mixer figure 4.14 gives an overview. For both mixing devices similar behaviour could be seen. After 6 min a sufficient dispersion was reached, while fibres shortened to a l_w value of about 2 mm. Further mixing lead to higher fibre breakage levels.

In a next step the weighted fibre length l_w was determined after complete processing. This means according to figure 3.2 incorporating the fibres either on two roll mill or internal mixer plus second step orienting the fibres and mixing in the curatives on two roll mill. Even by maintaining same processing conditions the scattering for the second step was significant higher, because of the subjective operator influence for two roll mixing.

The influence of the initial fibre length for the different dipped fibre types in EPDM on the l_w after processing step 1 and 2 is shown in figure 4.15. The differences after mixing in internal mixer vanished after second processing step, especially for Twaron. The fibres ended up with fibre lengths around 1 mm, irrespective of the initial fibre length. In case of RFL dipped Twaron the fibres could withstand the mixing procedure better, because of the filament agglomerates caused by the dip procedure. For Technora the additional breakage was very low.

The dependence of l_w values after both processing steps regarding the different matrices and fibre types used is displayed in figure 4.16. Another type of figure was introduced, because the datapoints began to overlap when using the type of figure 4.15. The black bar shows

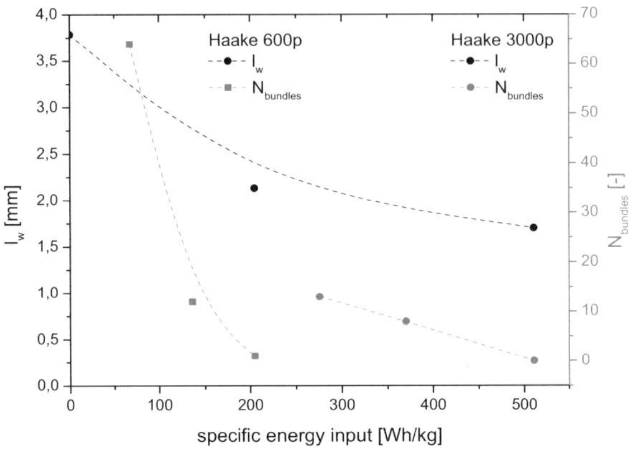

Figure 4.13: $N_{bundles}$ and l_w versus specific energy input for Twaron mixed in EPDM on Haake 610p or Haake 3000p internal mixer

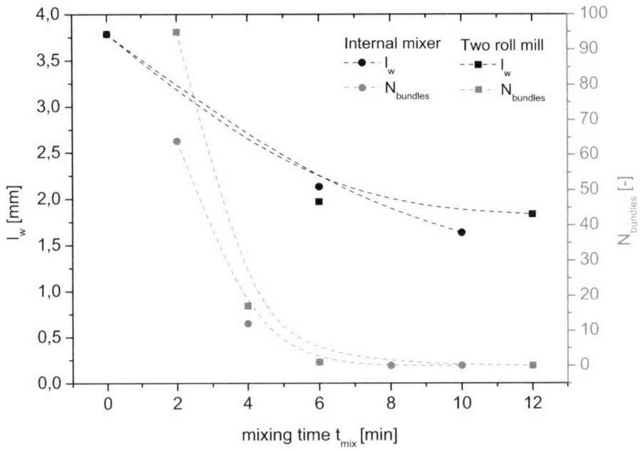

Figure 4.14: $N_{bundles}$ and l_w versus total strain for Twaron mixed in EPDM on Haake 610p internal mixer or two roll mill

hereby the weighted fibre length l_w after processing step 1, while the grey one represents processing l_w after step 2. Twaron halved the values for remaining l_w in EPDM compared to first mixing step, while Technora was nearly unaffected. In NR the fibre breakage of both fibre types was quite high. Technora dropped to a l_w of 1,1 mm while Twaron was even shorten to 0,4 mm. The reason for the significant high break was the mastication behaviour of the NR. Through the mastication process especially the high molecular weight polymer

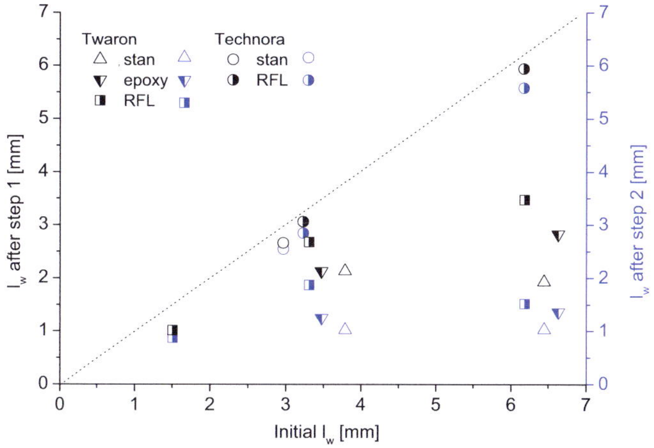

Figure 4.15: Influence of initial fibre lengths and surface treatments on l_w values after mixing step 1 on internal mixer and mixing step 2 on two roll mill for EPDM model compound

chains degrade and so the molecular weight and viscosity is going down [161]. Because of the applied two-step mixing process, the fibres were shorten to such high extent during the occurring high shear forces in the mastication phase on the two roll mill.

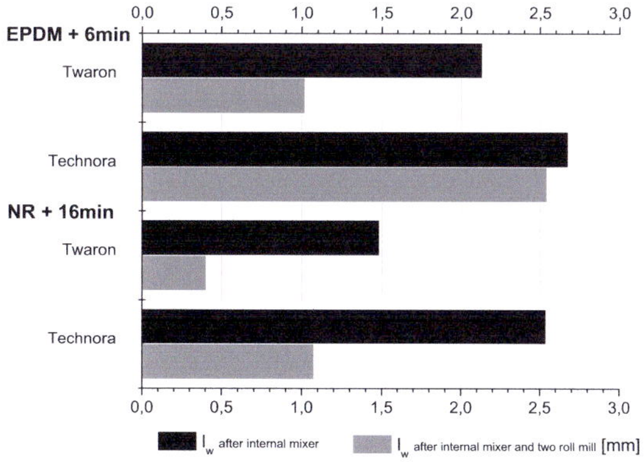

Figure 4.16: Influence of matrix and fibre type on l_w after mixing step 1 on internal mixer and mixing step 2 on two roll mill

Figure 4.17 compares the results after processing step 1 and 2 regarding the mixing devices used for incorporation of the fibres. Weighted fibre length l_w decreased to values of 1 mm for fibres mixed on internal mixer and incorporating the curing system on two roll mill. Contrary to that the fibres mixed in on two roll mill and mixing in curatives afterwards kept the lengths between 1,7 and 2 mm. These surprising results can be attributed to a higher predamage of the fibres in internal mixer, which is then amplified by the two roll mixing. Rajeev et al. [24] reported similar behaviour for mixing of melamine fibres in EPDM. However, for 5 phr mixed on two roll mill no decrease of l_w was observed. The values were even slightly higher. This can be attributed to difficulties by detecting so long fibres with at the same time such high fibre content. Therefore the fibre-fibre overlap was higher (see figure 4.22), which led to higher error in fibre length detection.

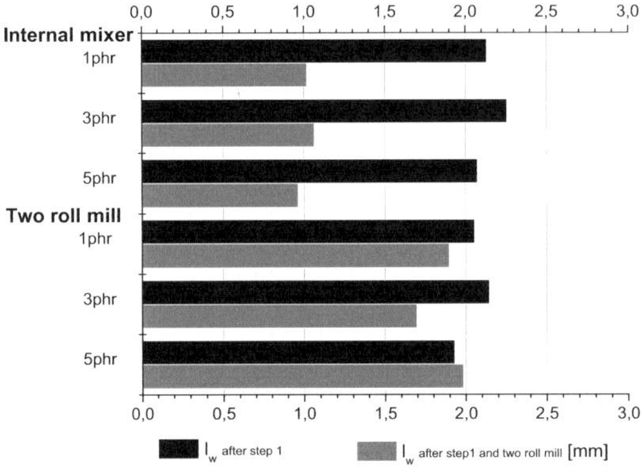

Figure 4.17: Influence of mixing device in processing step 1 on l_w after both processing steps

The dependence of l_w after both processing steps regarding the upscaling experiments on the internal mixers is displayed in figure 4.18. For Twaron an additional decrease of l_w values under 1 mm could be confirmed. In case of Technora values did not show further significant fibre breakage.

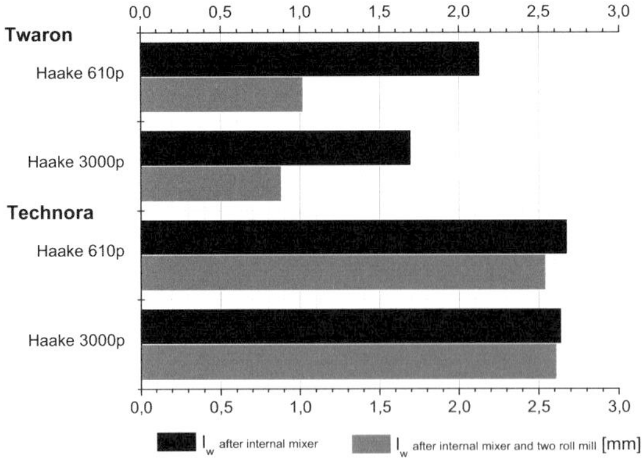

Figure 4.18: Influence of internal mixer size of processing step 1 on l_w after both processing steps

Irradiation of aramid fibres is a suitable procedure to increase the fibre-matrix interaction as reported in section 2.4.1, but at the same time aramid fibres are relative sensitive to irradiation. To increase the fibre-matrix interaction both fibre types Twaron and Technora were irradiated with doses of 10 or 20 kGy by the help of electron beam accelerator ELV-2 from Budker Institute of Nuclear Physics, Novosibirsk, Russia. According to Poncin et al. [98] 20 kGy is the optimum dose to create radicals for a postgrafting reaction on the aramid surface.

Radiation induced radicals were detected by Electron Paramagnetic Resonance Spectroscopy (EPR) with a MiniScope MS200 from Magnettech Limited, Berlin, Germany. Non-irradiated and irradiated fibres were analysed. The results are shown in figure 4.19. The zero

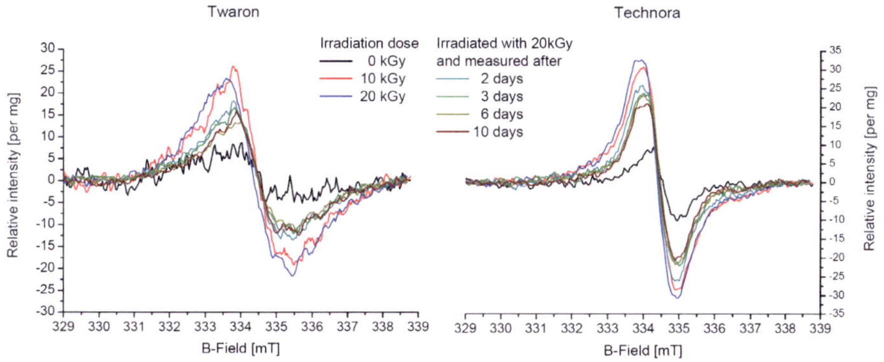

Figure 4.19: Electron Paramagnetic Resonance Spectroscopy graph for irradiated Twaron and Technora fibres with variation of dose and decay time

signal of non-irradiated fibres could be related to UV-light, inorganic paramagnetic ingredients from surface treatment, and fibre processing. The reduction of the relative intensity for radiation induced radicals after 9 days for Technora was about 35% and for Twaron about 45%.

The determination of the absolute number of radicals was carried out for Twaron and Technora. Both fibres generated an EPR spectrum consisting of different components which might result from a superposition of at least two several radicals. The spectra for the different fibres were not similar which could be originated by anisotropic spin distribution. This was caused by the anisotropic structure of the fibres (section 2.3.1.2). Therefore, a separation in single spectra and determination of the type of radical was not possible. To create an isotropic sample with homogenous distribution of the spins powder like samples would be necessary. With the actual sample shape the total error for the absolute number of radicals amounts was about 50%. The detection limit was 10^{14} radicals per g if the sample weight amounts to about 200 mg. Due the fibre geometry sample weights of 24 to 64 mg could be used for the measurements.

Also fibre length detection was executed for the irradiated fibres and results are shown in figure 4.20. When irradiating the Twaron fibres with a dose of 20 kGy, the l_w dropped dramatically to 1,3 mm. Any positive effect of improved fibre-elastomer interaction would have been superimposed by the decrease of fibre length during mixing. Contrary to that Technora did not show any significant enhanced fibre breakage. However, since the irradiation

Dose	$t_{irradiation}$	Twaron	Technora
kGy	days	10^{15} radicals per g	
0	0	2,7	0,6
20	0	4,8	2,8
20	1	3,1	2,1
20	2	2,8	1,9
20	5	2,5	1,8
20	9	2,6	1,8

Table 4.6: Number of radicals per g after application of a dose of 20 kGy as function of time

treatment showed considerable higher length reduction levels for the Twaron fibres after processing this method was not further pursued.

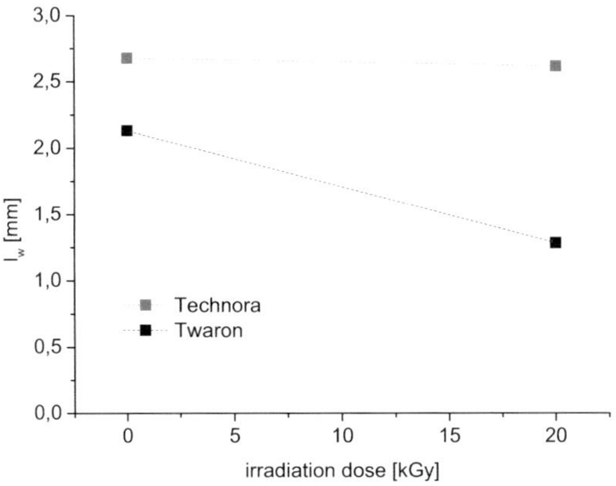

Figure 4.20: Effect of irradiation dose on l_w after mixing on internal mixer

4.3.2 Fibre diameter distribution

The determination of the fibre diameter distribution was interesting to investigate, since films of transparent model compounds studied for fibre length detection also showed differences in the fibre diameters regarding dips and fibre types as shown in figure 4.7. Due to the distortion of the fibre diameter dimension by light scattering, 30 objects with filament thickness were measured and the average was then normalised with 12 μm. Figure 4.21 shows the weighted fibre diameter distribution for EPDM and NR for a mixing time of 6 min in internal mixer.

Below the average fibre diameter interval of 10 to 15 μm fibrils with diameter between 0 to 10 μm could be found. These fibrils showed lengths up to about 500 μm and originated from the surface of fractured filaments. These kind of fibrils were reported by several authors [71, 75, 85] and are caused by the buckling susceptibility of the fibres.

For the RFL coated Twaron fibres also objects above the normal diameter, in the interval of 20 to 25 μm and 35 to 40 μm, were observed. These already in figure 4.7 (page 57) reported agglomerates or lumps were consisting of typically two or three filaments, which were glued together by RFL dip. These agglomerates are originating most probably from the dip procedure as reported in figure 4.1 (page 50). Also in NR these Twaron RFL agglom-

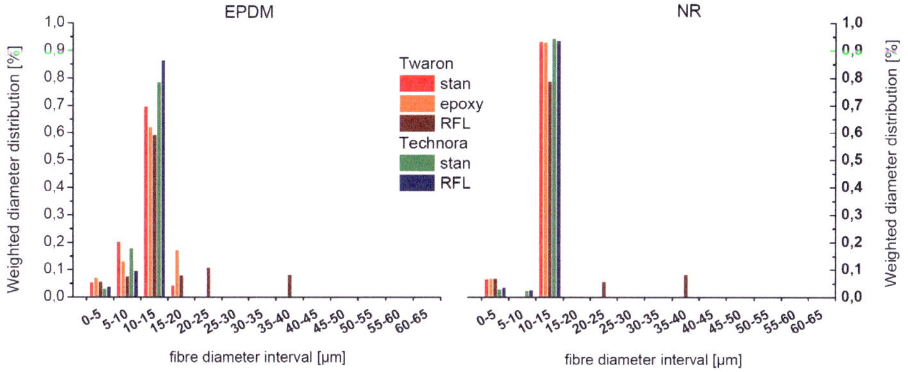

Figure 4.21: Diameter distribution for fibres after mixing 6 min in EPDM or NR

erates were observed. Contrary to EPDM the fibre diameter distribution appeared sharper below the nominal diameter fibre interval of 10 to 15 μm. The reason was that the optical transparency of the investigated matrices was different. While in EPDM the fluorescence intensity of fibrils was enough to be detected, in NR the lower matrix transparency hampered the detection of these fibrils.

4.3.3 Fibre-fibre interaction

To investigate possible differences between the fibre-fibre interactions regarding the two used mixing devices the fibre surface was detected for 1 and 5 phr Twaron fibres with standard finish in EPDM model compound. Therefore grey value images were generated for the 15 μm thick films used for fibre length detection by applying a threshold and the fibre surface was marked green (figure 4.22 and 4.23). It is clearly observable that the fibre-fibre interactions were increased for a higher fibre content of 5 phr leading to mechanical interlocking of the fibres. Figure 4.24 shows the detected fibre surface after processing step 2 regarding the

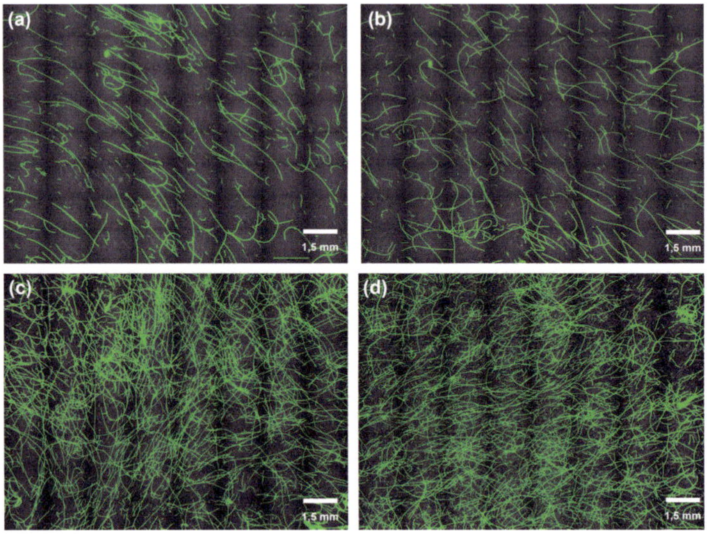

Figure 4.22: Images of fibre-fibre interaction for Twaron standard in EPDM after processing step 1: mixed in on internal mixer 1 phr (a) and 5 phr (c), and incorporated on two roll mill 1 phr (b) and 5 phr (d)

different mixing devices and fibre contents. There were two main differences compared to processing step 1. In case of 1 phr incorporated on internal mixer fibre length reduction as well as fibrils below the normal diameter of the filaments were observable. For 1 phr mixed on two roll mill the fibre length reduction was less pronounced, but there were more fibrils visible. With growing fibre content the described trends were amplified. The determined fibre surface in green was related to the overall image surface. The results can be seen in figure 4.24. Generally the percentage of detected fibre surface was much higher than the vol% reported in table 3.2. The values after processing step 1 regarding the different mixing devices and fibre contents were similar within the scattering. For 1 phr 6% and for 5 phr 18% could be detected. For 5 phr fibre content also the scattering went higher. After processing step 2 the fibre surface increased to 8% for 1 phr and for 5 phr to 23% in case of internal mixer and 19% for fibres incorporated on two roll mill. This confirms the mentioned different fibrillation behaviour comparing the images of figure 4.23 c and d.

Figure 4.25 shows the images of fibre interface area for 1 phr Twaron with standard finish in NR for mixing on internal mixer after processing step 1 (a) and 2 (b). The additional fibre

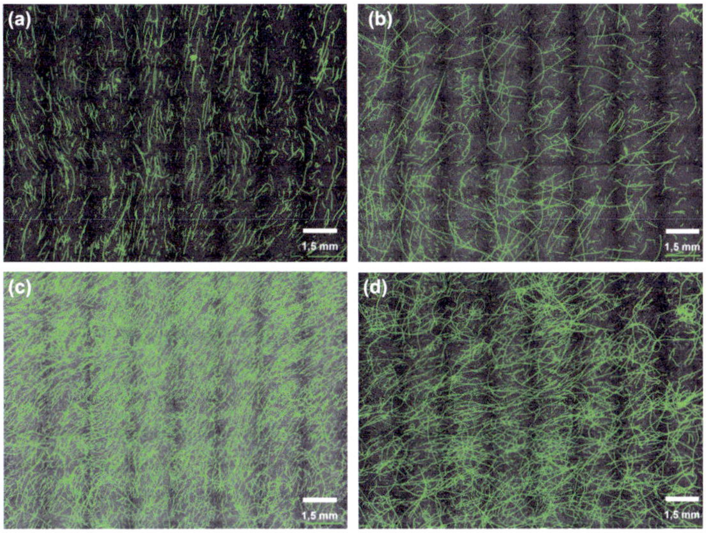

Figure 4.23: Images of fibre interface area for Twaron standard in EPDM after processing step 2: mixed in on internal mixer 1 phr (a) and 5 phr (c), and incorporated on two roll mill 1 phr (b) and 5 phr (d)

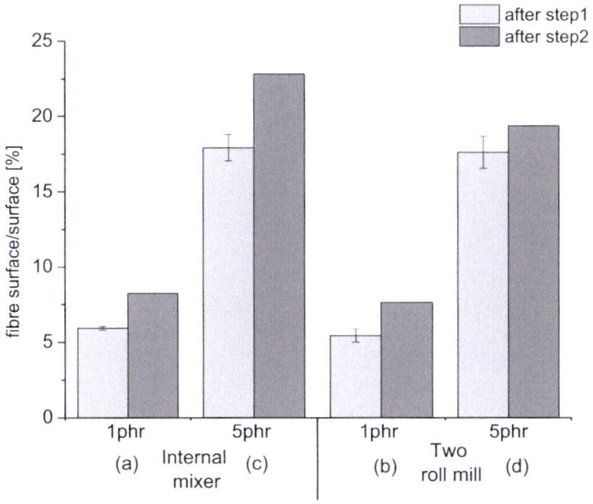

Figure 4.24: Detected fibre surface area in % for 1 and 5 phr Twaron standard fibre content after processing step 1 and 2

break was quite high which can be attributed to mastication behaviour of the NR matrix, with which each NR processing starts. If the fibres are already in the NR matrix before the mastication starts, as in the applied two-step mixing protocol, the fibres undergo high shear stresses during the initaly mastication mixing and therefore the breakage is enhanced.

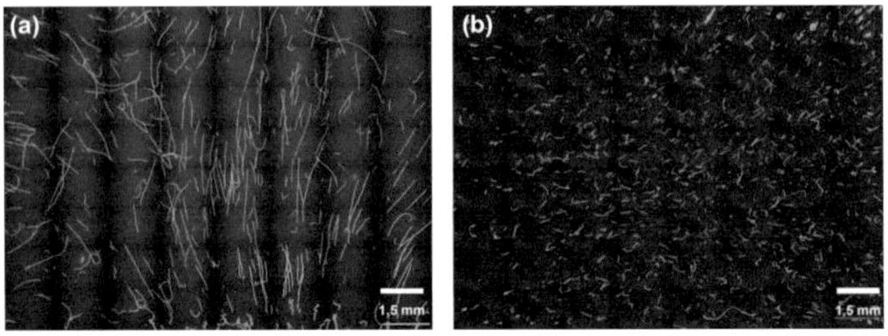

Figure 4.25: Images for fibre interface area for 1 phr Twaron standard in NR for mixing on internal mixer after processing step 1 (a) and step 2 (b)

To evaluate the fibre-fibre interactions in the compounds the minimum and maximum aspect ratios for 1, 3 and 5 phr were displayed in graph 4.26, where volume fraction is shown versus aspect ratio. According to Papathanasiou [10] fibre suspensions are classified into concentrated, semi-concentrated and dilute regions. Details can be found in section 2.1.2. For 1 phr or a volume fraction of 0,0058 the minimum value was obtained for Twaron mixed

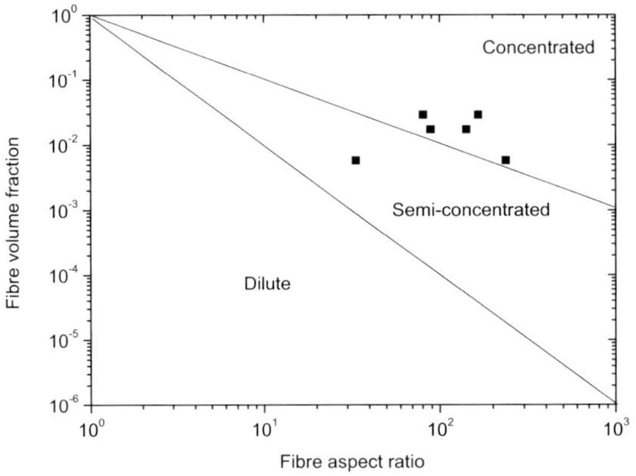

Figure 4.26: Fibre volume fraction versus fibre aspect ratio and corresponding minimum and maximum aspect ratios for 1, 3 and 5 phr

in 16 min in NR with an aspect ratio of 34. This point is visible in the semi-concentrated region, where the spacing between the fibres was less than their length but greater than their diameter. For 1 phr the limit between the concentrated and the semi-concentrated region was about 2 mm and between semi-concentrated and dilute 150 μm. Since Twaron was shortening under 2 mm this means that most of those compounds were in the semi-concentrated region, while for Technora they were in the concentrated region. All other fibre contents were lying in the concentrated region, where the spacing between the fibres was in the area of their diameter.

Summary

The aspect ratio was studied to investigate the influence of processing parameters like degree of fill, mixing energy input and mixing devices. Also material parameter like fibre content, initial length, type and surface treatment were studied.

While Technora fibres were relatively unaffected, Twaron fibre length breakage was dependent on the amount of specific energy input, total strain or mixing time. This was the reason for the direct influence of the initial length on the residual length for Technora, while for Twaron this was not a significant influence. For Twaron the RFL surface treatments had an effect on the length in the composite. Irradiating the fibres caused further fibre breakage in case of Twaron, while Technora could keep its length. Differences in Twaron fibre length after processing step one, the incorporation, were superimposed by the second processing step, the incorporation of curing additives and orienting of the fibres on two roll mill. Fibres processed only on two roll mill showed doubled residual length compared to those mixed on internal mixer. Since the fibres needed more time to disperse in NR and the high shear forces occurring during the mastication process, the fibre breakage was in general higher compared to EPDM.

Also the diameter distribution was changing during the mixing process. Below the average filament diameter which changed in the reported way, fibrils with a diameter ranging from 0 to 10 μm were observed. These fibrils showed lengths up to 500 μm. Also agglomerates containing multiple filaments could be observed in the case of Twaron with RFL dip.

In general the fibre-fibre interactions increased with growing aspect ratio or fibre volume content. Aspect ratio of the fibres with average filament diameter reached from 43 for Twaron with standard finish and 3 mm initial length mixed in NR until 465 for Technora with RFL dip and 6 mm initial length in EPDM. Therefore the fibre-fibre interactions were mainly belonging to the concentrated region or partly to the semi-concentrated region.

4.4 Fibre-elastomer interaction

4.4.1 EPDM and NR model system

SEM images of Twaron fibres after mixing showing the two main mechanisms for fibre size reduction for Twaron are displayed in figure 4.27. The left image shows a fibre with the kink bands, which were induced by the frequent shearing of the fibres during mixing. This kink bands caused cracks at the skin, which led finally to fibre breakage perpendicular to fibre axis (left image) and before that to skin fibrillation in fibre axis (right image) leading to separated fibrils as reported in section 4.3.2. The reason for this behaviour was the skin-core structure of the fibres (figure 2.5, page 14) and the resulting failure behaviour of the fibres. Figure 4.28 confirms the increasing Twaron fibre-fibre interaction for a growing fibre content

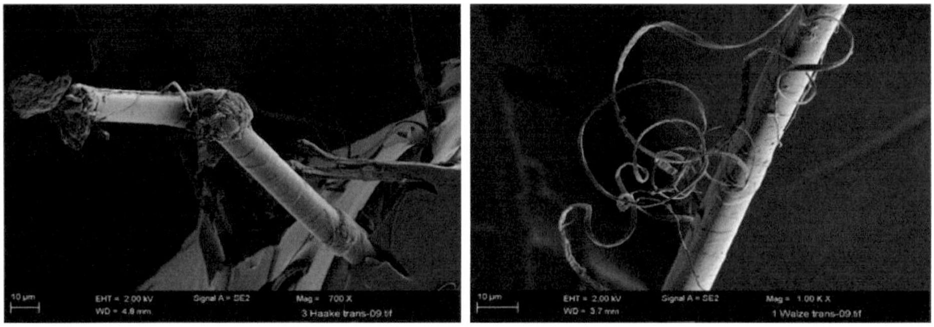

Figure 4.27: The two failure mechanisms: buckling break (left) and skin fibrillation (right)

of 1 phr (0,58 vol%) to 5 phr (2,86 vol%) for a specimen tested in fibre direction.

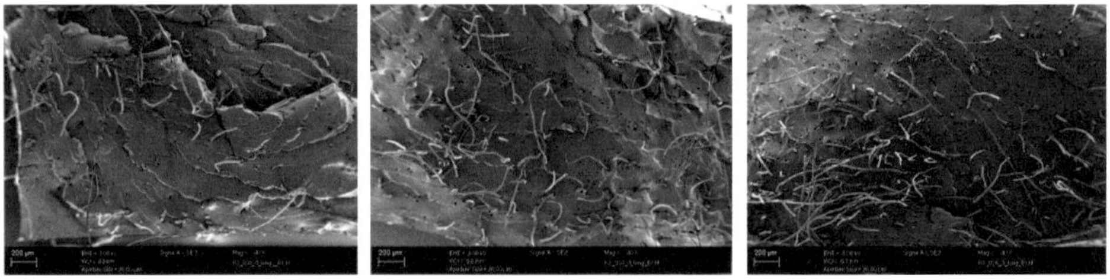

Figure 4.28: Longitudinal oriented Twaron fibres in EPDM model system: fibre content 1, 3 and 5 phr incorporated on Haake 610p (from the left to right)

Figure 4.29 displays the effect of fibre orientation at the surfaces of fractured tensile test bars for 5 phr Twaron standard in EPDM. Images taken from fractured tensile test bars showed that fibre rubber interface failed in form of pull-out holes in the longitudinal direction and in form of notches for the transversal oriented specimen. Lee [39] investigated SEM images of short glass fibre reinforced CR and stated that this adhesive failure mode is typical for composite materials having a soft matrix and hard fillers.

Representative images for fibre-matrix adhesion between RFL dipped fibres in EPDM and NR model systems are shown in figure 4.30. Also for the fibres treated with RFL dip

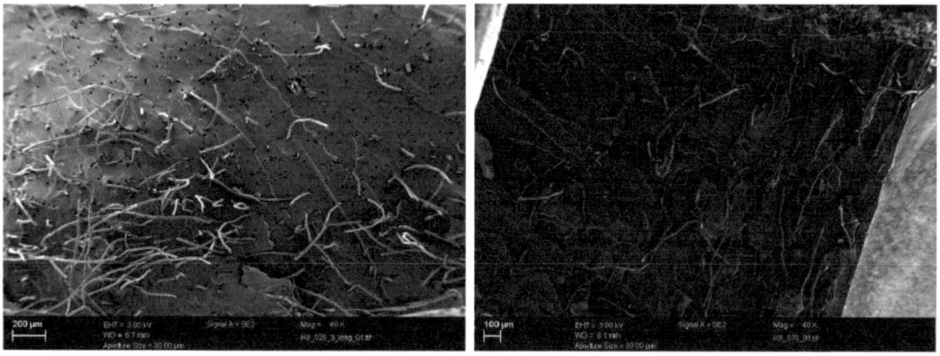

Figure 4.29: Images for effect of fibre orientation for 5 phr Twaron standard in EPDM: longitudinal (left) and transversal (right)

in both matrices the fibre-rubber interface failed, indicated by pull-out of fibres. If there were residuals on the fibre surface, it was unclear if they were originating from the matrix or the RFL finish itself. Clarke and Harris [116] stated that even with a high fibre-rubber bonding strength the fractures will be initiated at the high stress region between rubber and fibre. The detailed SEM images in figure 4.30 confirmed also the presence of agglomerates for Twaron fibres with RFL dip (a) also found in figure 4.7 (page 57) for dispersion studies. Image (b) shows that in case of Technora fibres fibrillate parallel to fibre axis, while for Twaron fibre fracture dominated (c).

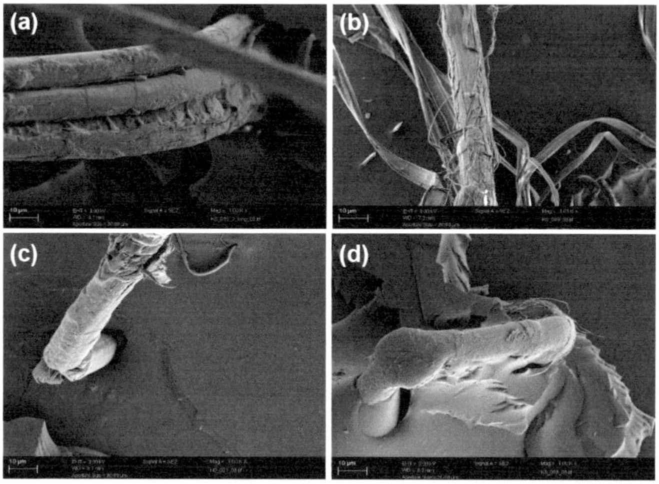

Figure 4.30: Images for fibre-elastomer adhesion: EPDM with Twaron RFL (a) and Technora RFL (b), NR with Twaron RFL (c) and Technora RFL (d)

4.4.2 EPDM and NR industrial systems

Figure 4.31 shows the increasing fibre-fibre interaction for the broken tensile test bars in fibre direction for EPDM hose and NR tread compounds in dependence of fibre content. Better orientation was observable for EPDM. Fibre-fibre interaction was increased for fibre contents of 5 phr (1,35 vol%).

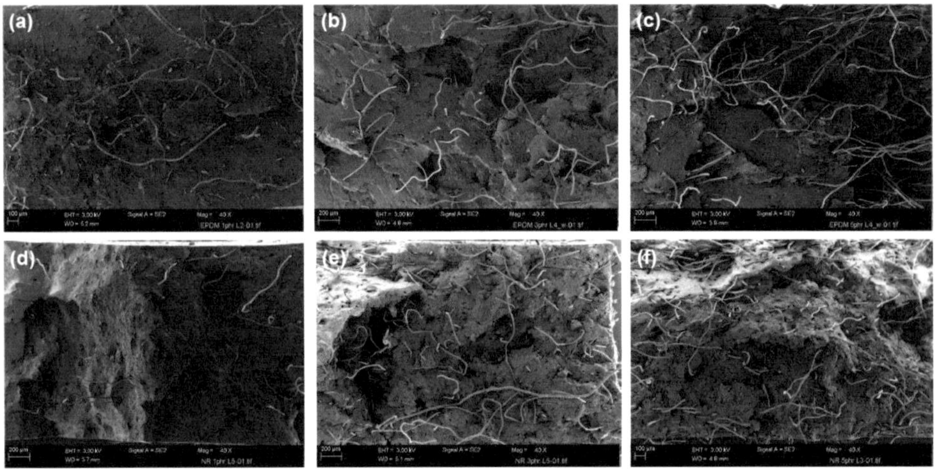

Figure 4.31: Longitudinal oriented Twaron fibres in EPDM/NR systems: EPDM 1 (a), 3 (b) and 5 phr (c); NR 1 (d), 3 (e) and 5 phr (f)

In next Figure 4.32 the effect of fibre orientation is shown. Tensile test bars in and perpendicular to fibre direction for EPDM hose and NR tread compounds are displayed for 5 phr Twaron fibre content. Similar to the model systems an insufficient fibre-elastomer interaction was indicated in fibre direction in form of pull-out holes and perpendicular to it in form of notches. Contrary to the model systems the orientation was less distinctive. In all images both indicators were present.

SEM images for fibre-matrix adhesion between RFL dipped fibres in EPDM and NR carbon black filled systems are shown in figure 4.33. Particles were found in all SEM images for EPDM hose compounds blooming out of the matrix in form of crystals on the surfaces of fractured tensile test bars. Especially image (a) shows these crystals adhering to the surface of the RFL dipped Twaron fibre. They were possibly originated by a solution problem of the peroxide system used [162]. In NR like expected no crystals were observable since a sulphur curing system was used. The effect of the found particles on the fibre-matrix interaction is unknown, since the particles just appeared on surfaces and may therefore not act in the composite.

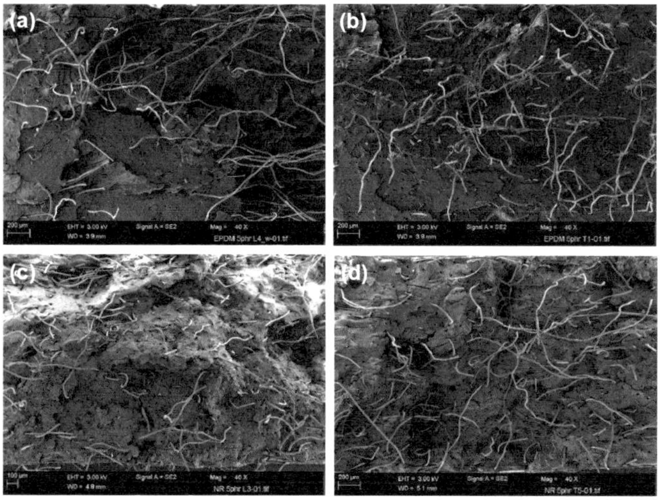

Figure 4.32: Images for effect of fibre orientation for 5 phr Twaron standard: In fibre direction in EPDM (a) and NR (c), perpendicular to fibre direction in EPDM (b) and NR (d)

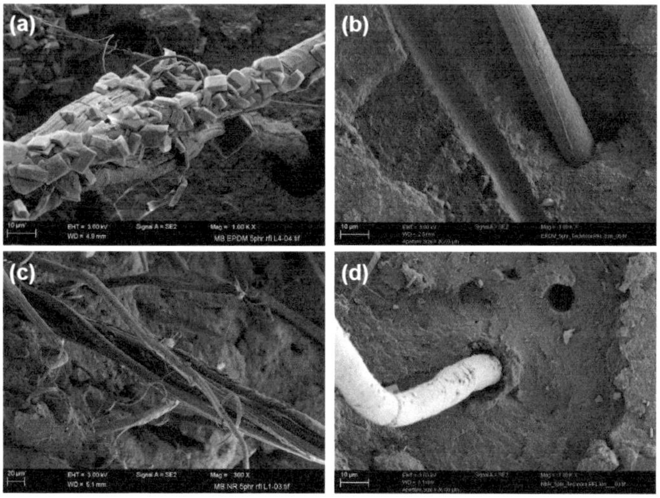

Figure 4.33: Images for fibre-elastomer adhesion: EPDM with Twaron RFL (a) and Technora RFL (b), NR with Twaron RFL (c) and Technora RFL (d)

4.4.3 Critical fibre length

It is possible to calculate a critical minimum fibre length, which is required for obtaining complete stress transfer assuming a perfect bond between the fibre and the matrix, based on the equation proposed by Rosen [15]:

$$l_c = \left[\frac{1}{2} \times \frac{E_f}{G_m} \times \frac{1 - \phi_f^{1/2}}{\phi_f^{1/2}} \right]^{1/2} \cdot d \tag{4.1}$$

Table 4.7 shows the calculated l_c values for the investigated compounds according to input values in table 6.1. Due to the slightly differences between the elastic fibre moduli E_f of Twaron and Technora compared to the matrix shear modulus G_m, the l_c is given in full millimetre to display it for both fibre types at the same time.

The calculated l_c values for the EPDM model and hose compound were similar. For 1 phr fibre content it was determined with 8 mm. This means that the occurring fibre lengths for 1 phr (with residual lengths up to 6 mm) were shorter than the critical length. In theory that would mean that not full fibre strength was utilised and that bond strength is critical. Therefore the fibre-elastomer bond or the elastomer matrix itself will fail before the ultimate fibre strength is reached. This will lead to fibre pull-out from the matrix, which was already confirmed by the SEM images (table 4.7). However, above 3 phr the l_c dropped under 6 mm for most investigated compounds and were therefore in the range of the fibres with corresponding residual lengths.

For NR model the low G_m led to a increase of fibre lengths over 7 mm even for 5 phr. Since the NR tread obtained higher G_m values the opposite trend could be observed. Even for 1 phr l_c dropped under 6 mm.

Content l_c	[phr] [mm]	1	3	5
EPDM	Model	8	6	5
	Hose	8	6	5
NR	Model	10	8	7
	Tread	5	4	3

Table 4.7: Calculated l_c values for investigated compounds

Summary

Looking at SEM images of the fracture surfaces of the tested tensile specimen it became clear, that fibre-elastomer interaction was mainly caused by mechanical interlocking due to the roughness of the fibre surface. The calculated critical fibre length could not be reached since it was exceeding the initial fibre length. Therefore the fibres were pulled out of the matrix before the full fibre strength was reached. However, since the strength of the matrix was magnitudes under that of the fibre the failure of the composite was initiating either in the matrix or in the fibre-matrix bond which is accordance with other works [16,116]. Beside this findings also the two main fibre breakage mechanism during mixing, namely buckling break and skin fibrillation could be confirmed as well as the trends for main orientations and fibre-fibre interactions.

4.5 Curing characteristics and Mooney viscosity

4.5.1 EPDM and NR model system

Characteristic curing values like minimum- and maximum torque $S'min$, $S'Max$, $\Delta S'$, characteristic curing times t_{10}, t_{90}, t_{90}-t_{10} are displayed for EPDM model system in table 4.8. In

	S' Min [dNm]	*S' Max* [dNm]	$\Delta S'$ [dNm]	t_{10} [min]	t_{90} [min]	$t_{90} - t_{10}$ [min]
EPDM K8340	0,45	4,80	4,35	1,35	15,44	14,09
Twaron						
1 phr stan	0,50	5,34	4,84	1,27	15,54	14,27
1 phr epoxy	0,53	5,49	4,96	1,22	13,90	12,68
1 phr RFL	0,50	5,37	4,87	1,18	13,52	12,34
3 phr stan	0,61	6,44	5,83	1,22	14,09	12,87
5 phr stan	0,79	6,65	5,86	1,21	13,74	12,53
Technora						
1 phr stan	0,43	4,91	4,48	1,30	13,80	12,50
1 phr RFL	0,42	5,23	4,81	1,25	14,26	13,01
EPDM K2340	0,11	4,05	3,94	1,62	17,13	15,51
Twaron 1phr stan	0,13	4,56	4,43	1,62	16,77	15,15
EPDM 50-50 mix	0,25	4,59	4,34	1,35	15,08	13,73
Twaron 1phr stan	0,28	4,93	4,65	1,42	15,72	14,30

Table 4.8: Curing values for EPDM model system

general the fibre orientation was assumed to be planar random in the testing chamber.

The $\Delta S'$ values went higher with increasing fibre content, while the curing times t_{10} and t_{90} decreased. Similar behaviour was found for the curing of short oil palm fibre reinforced NR [60]. The increase of the torque values is a result of the increased viscosity, due to the hydrodynamic reinforcement of the fibres in the matrix. This effect is more pronounced when the shear rates are relatively low. If the shear rates are increasing, like for extrusion processes, the viscosity is going down showing the so called pseudoplastic behaviour [41].

The decrease of optimum curing time t_{90} with increasing fibre content for both NR and EPDM model and industrial compounds is caused by the higher thermal conductivity of the compounds. This effect leads then to a faster heat transfer through the fibre-elastomer interfaces into the matrix and more shear induced heating, which is causing lower times required to start the vulcanisation process. The same effect was discussed by Ürögiova et al. [163] for metal coated carbon black and by Manchado et al. [25] for short fibres in various elastomer matrices. In case of the different fibre dips for constant fibre content the curing characteristics were similar and no clear trends were observed.

The low viscosity EPDM type K2340 and the 50-50 mix showed generally lower $S'min$ and $S'Max$ values for the gum and fibre filled compounds. However, the trends regarding the fibre addition were similar to those of the compounds with EPDM type K8340.

Since for NR no fibre content variation could be carried out due to dispersion problems, only the curing values for different fibre types and treatments are shown in table 4.9. The values did not show significant differences, regarding fibre type or surface treatment. Generally the maximum torque levels, $S'Max$, were increased compared to fibre free compound. Also the optimum curing time t_{90} was slightly declining, similar to the findings for the EPDM model compound. In contrast to EPDM model compound the t_{90} values were lower since much higher concentration of DCP was used for curing.

	S' Min [dNm]	S' Max [dNm]	$\Delta S'$ [dNm]	t_{10} [min]	t_{90} [min]	$t_{90} - t_{10}$ [min]
NR	0,14	1,68	1,54	0,79	8,89	8,10
Twaron						
1phr stan	0,05	2,16	2,11	0,83	7,28	6,45
1phr epoxy	0,09	1,86	1,77	0,76	6,50	5,74
1phr RFL	0,11	1,82	1,71	0,77	6,49	5,72
Technora						
1phr stan	0,11	1,87	1,76	0,78	7,99	7,21
1phr RFL	0,17	2,09	1,92	0,71	7,10	6,39

Table 4.9: Curing values for NR model system

4.5.2 EPDM and NR industrial compounds

The measured and calculated curing values of the EPDM hose compounds are observable in table 4.10 together with the Mooney viscosities. Similar to the model compounds an increase in torque level and a decrease of curing times values could be seen. For higher fibre contents the rise of fibre content was not affecting the values so remarkable as it could be shown for the model systems. This was caused by the presence of plasticizers like oil and stearic acid and the lower effective vol% reported in table 3.2 (page 34). For the different fibre types and dips no clear trend could observed. The t_{90} for the specimen filled with 5 phr Technora fibres with standard finish was unexpected high, even exceeding that of the fibre free matrix. Mooney viscosity was just slightly increasing with growing fibre content.

	Mooney [MU]	S' Min [dNm]	S' Max [dNm]	$\Delta S'$ [dNm]	t_{10} [min]	t_{90} [min]	$t_{90} - t_{10}$ [min]
EPDM hose	60	0,50	5,34	4,84	0,99	10,56	9,57
Twaron							
1phr stan	62	0,55	5,62	5,07	0,88	9,61	8,73
3phr stan		0,54	5,69	5,15	0,95	9,43	8,48
5phr stan	62	0,61	5,98	5,37	0,95	8,96	8,01
5phr epoxy	62	0,55	7,07	6,52	0,91	8,98	8,07
5phr RFL	63	0,52	6,15	5,63	0,98	8,85	7,87
Technora							
5phr stan		0,46	6,10	5,64	1,10	11,12	10,02
5phr RFL		0,44	5,96	5,52	0,99	8,53	7,54

Table 4.10: Curing values for EPDM hose compound

Table 4.11 shows the curing values for the NR tread compounds. Similar to the EPDM hose compounds the torque values and Mooney viscosities were increasing while the curing times declined with growing fibre content.

	Mooney [MU]	S' Min [dNm]	S' Max [dNm]	$\Delta S'$ [dNm]	t_{10} [min]	t_{90} [min]	$t_{90} - t_{10}$ [min]
MB NR	47	0,80	6,89	6,09	3,70	25,25	21,55
Twaron							
1phr stan	48	0,63	7,39	6,76	5,81	22,21	16,40
3phr stan		0,81	7,92	7,11	7,24	23,52	16,28
5phr stan	49	0,58	8,24	7,66	1,77	23,46	21,69
Technora							
5phr stan		0,31	8,29	7,98	3,47	18,94	15,47
5phr RFL		0,35	7,99	7,64	2,81	19,58	16,77

Table 4.11: Curing values for NR tread compound

Summary

In general the influence of the different fibre types and surface treatments applied was not significant. To summarise the curing behaviour regarding the fibre content figure 4.34 shows the normalised maximum torque $S'Max$ and optimum curing time t_{90} values for the studied model and carbon black filled compounds in dependence of fibre content in vol%. A similar increase of $S'Max$ could be observed for both, model or industrial compounds, while for t_{90} the decrease followed no clear trend.

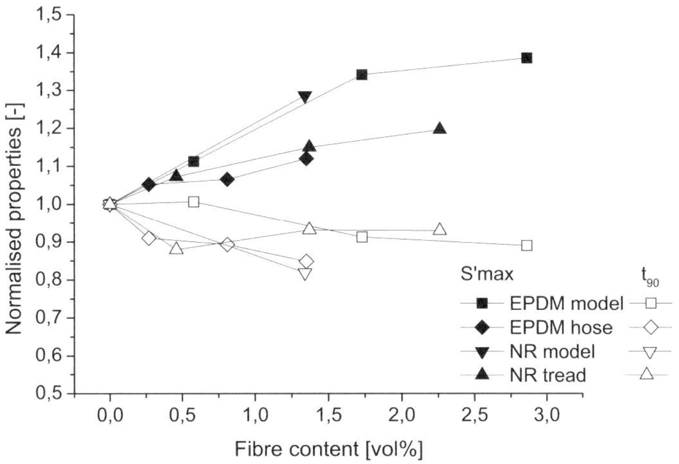

Figure 4.34: Comparison of normalised curing values $S'Max$ and t_{90} for the NR and EPDM model and industrial systems versus of fibre content in vol%

4.6 Fibre orientation distribution

4.6.1 Sampling in the mould

Do determine the influence of the flow induced orientation change in the mould five tensile test specimens were punched out of the mould in regular distances observable in figure 4.35. Confocal laser scanning microscopy was carried out for number 1, 3 and 5. This positions had corresponding distances to overflow channels x of 10, 50 and 20 mm. Position number 1 and 5 were influenced by the flow of the overflow channels, while 3 should not be affected. Specimens for dynamic mechanical analysis (DMA) were punched out between tensile sample 3 and 4 to achieve optimal fibre orientation.

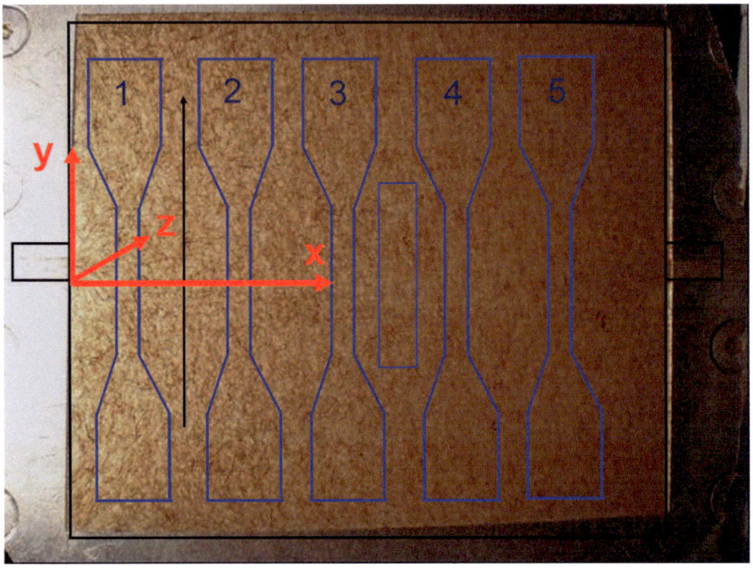

Figure 4.35: Vulcanised rubber sheet with numbered sampling positions

4.6.2 Orientation in dependence of position in the mould

The main orientation and its deviation angle from the direction of tension in the tensile test bars at position 1, 3 and 5 were determined. This angle α and the orientation factor o_f in dependence of the position in the mould are shown in the following figures together with best oriented CLSM planes for several influence factors. It is important to state that just for the precondition of a high o_f value, a low α is important to influence the resulting orientation. If the o_f value is too low, the influence of α value can be neglected, since the fibres are not oriented and α is an average of a random orientation only.

The dependence of o_f and α for increasing z-direction of scanned CLSM planes is shown in figure 4.36 for one example. While with increasing z-direction the orientation factor o_f is slightly scattering on high level, the deviation angle α is remaining on low level. To compare the different morphologies CLSM planes with a z-depth of $100\,\mu m$ were investigated in the following to achieve the best compromise between morphology visibility and comparability between EPDM and NR, because the matrices had different laser light loss over the z-depth.

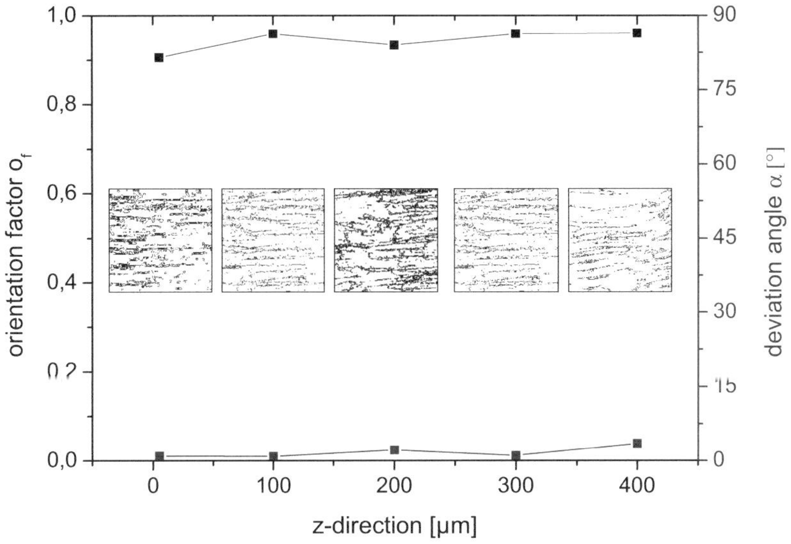

Figure 4.36: o_f and α dependency on increasing z-direction of scanned CLSM planes

The following figure shows the influence of Twaron fibre content on o_f and α regarding the different mixing devices and the corresponding CLSM planes (figure 4.37). The α values for all investigated samples showed a minimum in the middle of the mould at position 3, where the flow into the overflow channels was not affecting the orientation anymore. Regarding the o_f values differences were found. While for 1 phr internal mixer sample the highest value was detected at position 1 with 0,9, the other samples had their maximum at position 3 as expected. Irrespective of the mixing device the 5 phr specimen had a higher o_f level compared to the 1 phr samples. This can be explained by the higher fibre-fibre interactions in the elastomer, which lead to a higher fibre alignment. Especially for 5 phr mixed on two roll mill this was the case, since the fibres were significantly longer compared to those of the specimen mixed on internal mixer. This confirms the findings for the fibre length distribution in figure 4.17 (page 64).

Figure 4.38 shows the corresponding CLSM planes and the o_f and α values versus the position in the mould for Twaron or Technora with standard or RFL dip incorporated on internal mixer in EPDM. The agglomerates described in section 4.2 as well as the findings that RFL dipped Twaron fibres were longer than standard dipped and that Technora fibres generally shorten less were confirmed by the CLSM plane images. Figure 4.38 d also shows the buckling points induced by mixing already observable in figure 4.27 (page 72). All specimens

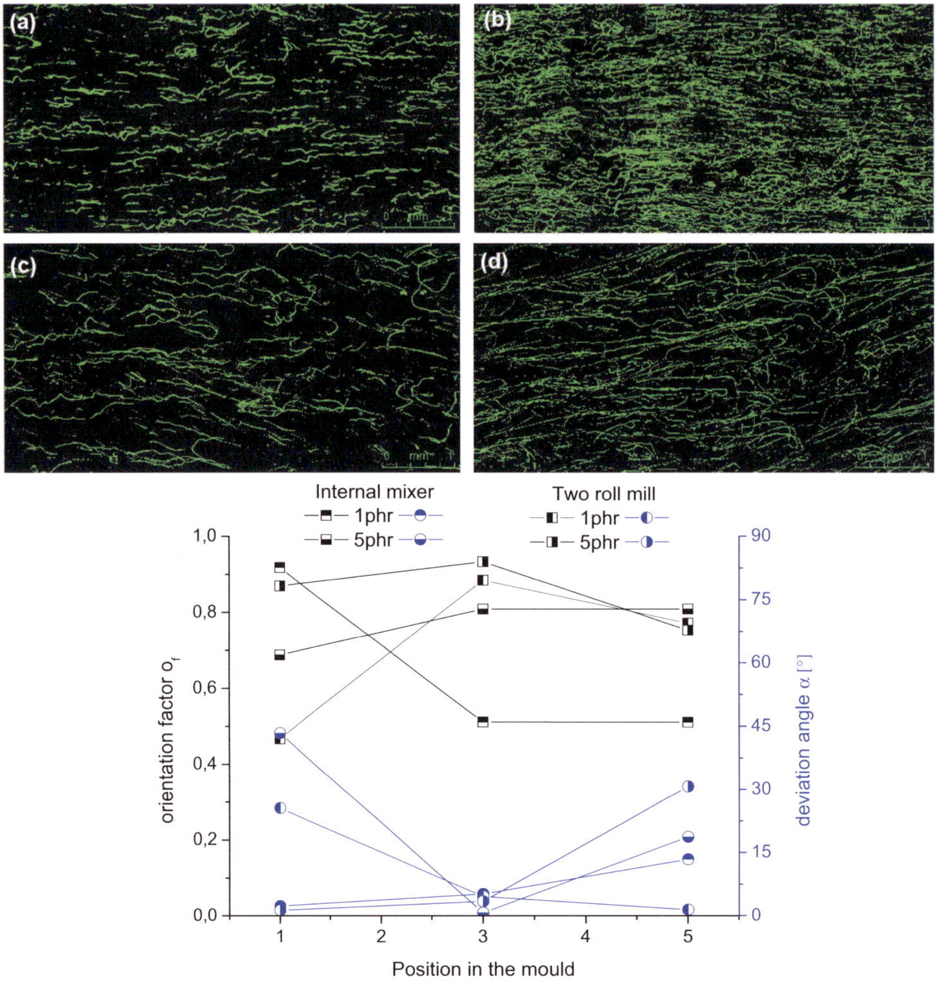

Figure 4.37: CLSM planes with best orientation for Twaron with standard dip incorporated in EPDM with internal mixer 1 phr (a) and 5 phr (b) or two roll 1 phr (c) and 5 phr (d) and corresponding o_f and α versus position in the mould

showed again a minimum for α at the middle of the mould at position 3, while o_f values went through a maximum at this position. As mentioned this was attributed to the distance of the overflow channels. One exception of this trend was the Twaron sample with standard dip, which showed highest orientation at position 1. Technora dipped with RFL showed the highest o_f and at the same time lowest α values for all positions indicating the highest average main orientation of the investigated specimen.

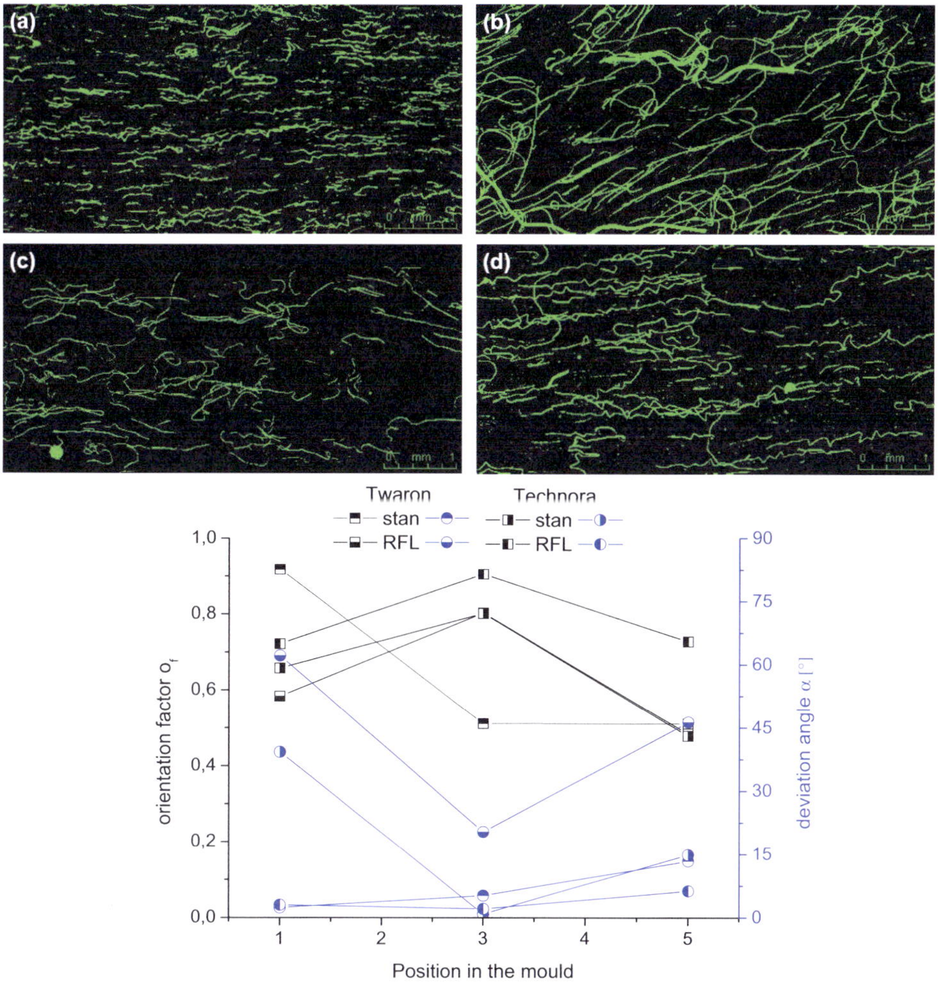

Figure 4.38: CLSM planes with best orientation for fibres in 1 phr EPDM Twaron standard (a) or RFL (b), Technora standard (c) or RFL (d) and corresponding o_f and α versus position in the mould

The influence of the different fibre types with standard dip oriented in EPDM or NR matrices is displayed in figure 4.39. The found fibre shortening due to the higher mixing times needed for dispersing fibres and the mastication in mixing step 2 in NR reported in figure 4.25 (page 70) could be confirmed by the CLSM planes.

The mentioned trends of o_f maxima and α minima at position 3 at the middle of the mould with the exception of Twaron 1 phr were confirmed. Generally the differences between o_f values of the different matrices were not large, but the maxima for EPDM were higher compared to NR.

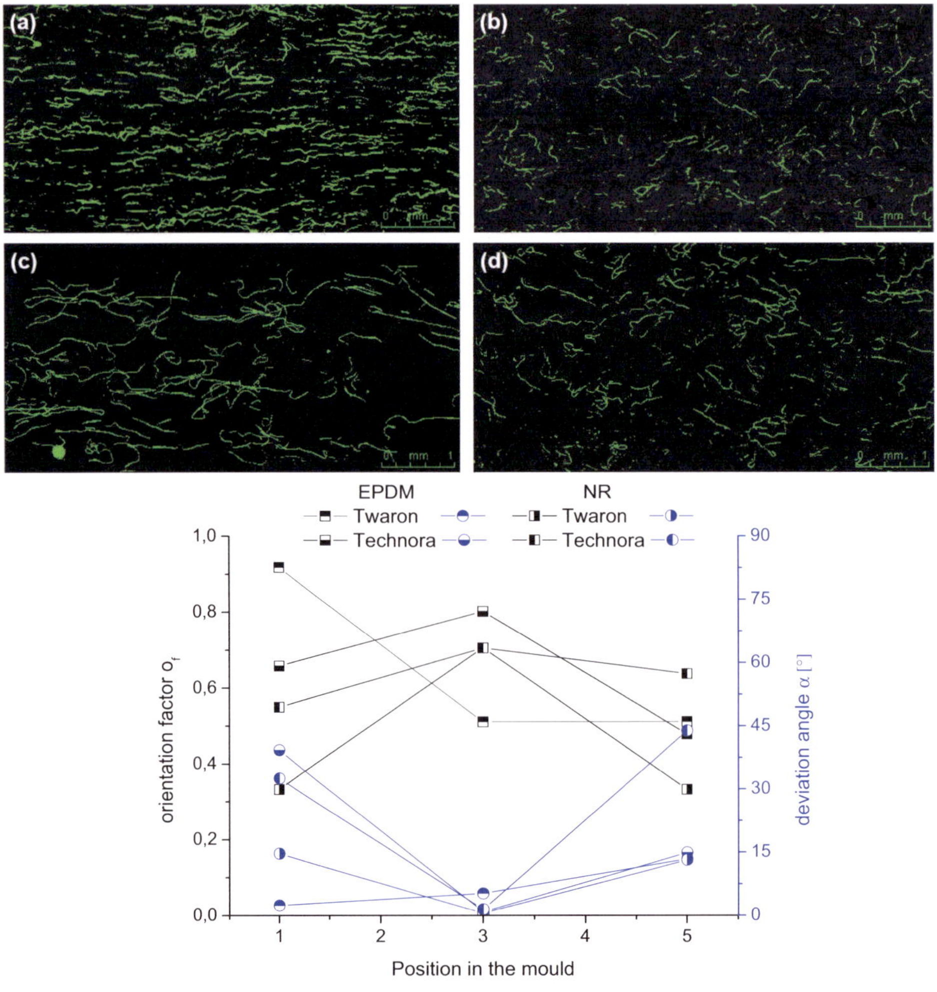

Figure 4.39: CLSM planes with best orientation for 1 phr Twaron standard in EPDM (a) or NR (b), 1 phr Technora standard in EPDM (c) or NR (d) and corresponding o_f and α versus position in the mould

4.6.3 Orientation in rubber discs for tribological tests

To study the influence of different fibre orientations on the tribological properties, corresponding rubber discs were made at IPF according to section 3.2.3 and later tested at OTR. To investigate the press induced distortion of the desired orientations, transparent rubber discs with 1 phr RFL treated Twaron fibres were created, because the orientation was visible by eye due to the mentioned agglomerates figure 4.7 (page 57). Figure 4.40 shows the transmission light images of the three orientations from the left to the right:

- Normal (fibres oriented perpendicular to the tribological surface)
- Longitudinal (fibre orientated circumferential to disc axis and in plane with the tribological surface)
- Random

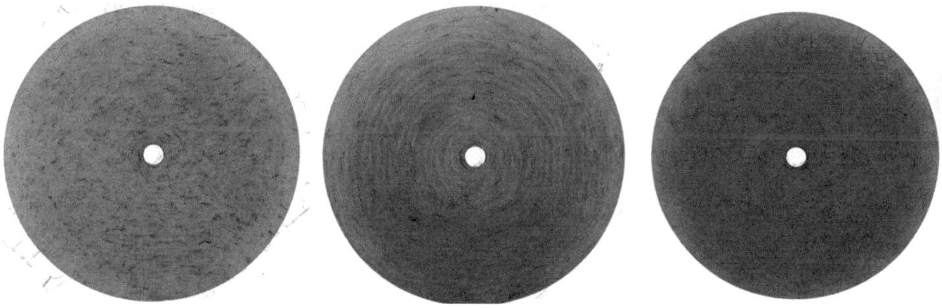

Figure 4.40: Rubber discs for EPDM model system with orientation from left to right: normal, longitudinal and random

The bulk orientation seemed to be in line. For the surface layer the rubber discs were cut and photos were taken. As discussed in section 2.6.3 the normal orientation should have the best abrasion resistance and coefficient of friction. Therefore the normal orientation was in focus, since the question appeared if the fibres were oriented till the tribological surface in the desired direction. The crosscut of the normal oriented EPDM model rubber discs can be seen in figure 4.41. It confirmed that the short fibres were orientated till the surface in normal direction. However, for the longitudinal orientation the Twaron RFL agglomerates would be a problem, since the clumps will be easily torn of the rubber matrix.

Figure 4.41: Cut of the normal oriented rubber disc

4.7 Tensile behaviour

4.7.1 EPDM and NR model system

In the following the representative stress-strain curves of five samples, if not else indicated, in fibre direction are displayed. The results of tensile testing are compared regarding different fibre contents, main orientations, mixing devices, processing conditions, upscaling behaviour, fibre types and surface treatments in different matrices.

Due to orientation of the fibres in different directions, it is possible to tune the resulting mechanical properties. The influence of fibre main orientation with respect to the fibre content is displayed in figure 4.42, which compares the tensile curves of longitudinal and transversal oriented fibres incorporated in EPDM on internal mixer. With growing fibre content the initial slopes for compounds oriented in fibre direction at low elongations got steeper, while strain at break values decreased. This was attributed to the increasing fibre-matrix interface with growing fibre content, which immobilised the rubber molecules and caused the increase of stress level in the elongation area under 50%. For 5 phr the drop after the yield point or maximum stress was highest, but the higher fibre content initiated cracks, which lowered the elongation of break dramatically.

Only for 1 phr Twaron in fibre direction the level of stress could be raised, without drop of strain at break value. The reinforcement could significantly observed till 50% strain. For higher strains the fibre rubber interface failed and the fibres were pulled out. A similar behaviour was found by Clarke und Harris [116] for cellulose filled NR vulcanisates or Naskar and Shibulal [54] for aramid filled ethylene octene copolymer. The transversal orientated fibres exhibited a low reinforcement, showing slightly increased stress levels between those of 1 phr longitudinal specimen and the gum reference and high elongation at strain. This was caused by the fact that perpendicular to fibre length just the diameter of the fibres was interacting with the matrix and not the full length.

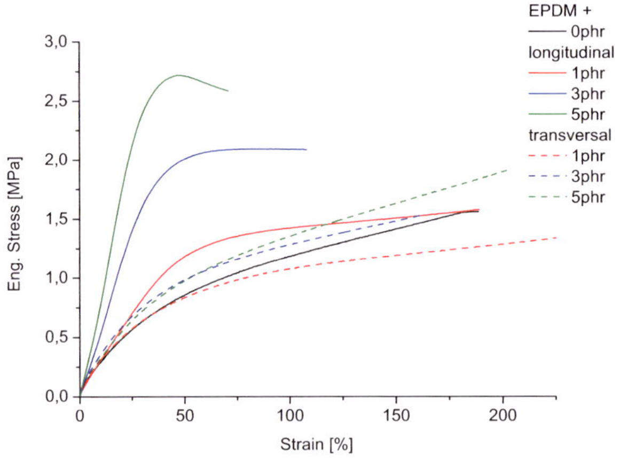

Figure 4.42: Comparison of longitudinal and transversal orientated fibres incorporated by internal mixer with varying fibre content

As mentioned in section 4.6.2 the orientation is also dependent on the distance to the overflow channel x and therefore from the position in the mould. Even when the fibres were originally well oriented, the overflow channels could influence the deviation angle α from the direction of tension and the orientation factor o_f significantly. Therefore in figure 4.43 the dependence of the stress-strain curves and orientation predictors α and o_f, and the measured Young's modulus from the position in the mould are displayed for the 5 phr Twaron standard mixed in EPDM on internal mixer. With getting nearer to position 3, the o_f values increased, while α declined. Therefore the measured Young's modulus had a maximum at position 3 and decreased when coming nearer to overflow channels. The corresponding tensile curves were showing the same trend.

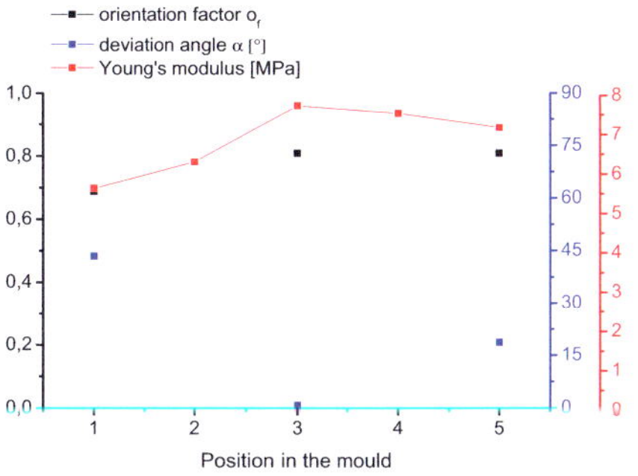

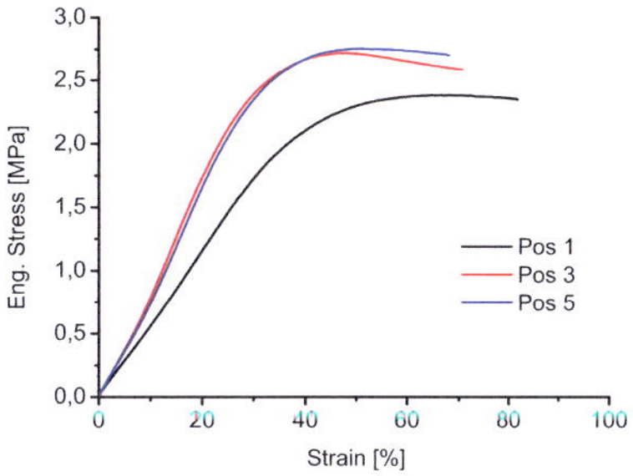

Figure 4.43: Dependence of α and o_f, and the measured Young's modulus (left) and stress-strain curves (right) from the positions in the mould for 5 phr Twaron

Figure 4.44 shows the influence of fibre content and mixing device on stress-strain behaviour of EPDM filled with Twaron standard treated fibres. As reported in section 4.3.1 the fibre length after complete processing for fibres mixed on internal mixer obtained half of the values of those mixed on two roll mill. The effect of higher fibre-matrix interface caused by the higher fibre length was enhancing with increasing the fibre content. So the maximum stress values and Young's modulus values were rising. Therefore Young's modulus for 5 phr processed on two roll mill was three times higher than the value for 5 phr processed on internal mixer.

The stress-strain curves for Twaron and Technora with standard dip regarding the mixer size is displayed in figure 4.45. Since the differences of fibre lengths after the first processing step (mixing in) were neutralised after the second processing step (orienting and mixing in of curatives, see figure 4.18 (page 64), also the differences between the corresponding tensile curves were not significant. Due to this overlay the differences for variations of conditions in the internal mixer in step one, did not affected the final tensile curves of the compounds.

The influence of initial fibre length, type and dip for 1 phr fibre content mixed in EPDM by internal mixer is shown in next figure 4.46. As shown in figure 4.15 page 63) Twaron had because of its higher affinity to fibre breakage an optimum initial fibre length of 3 mm to keep the highest residual lengths. This can be confirmed by the curves. Twaron fibres obtained a moderate reinforcement showing slightly increased stress levels by simultaneously little dropped strains at break. The higher initial length did not result in higher reinforcement, which was caused by the fibre breakage. This trend can be seen exemplary for Twaron fibres with RFL dip. For 1 mm the initial slope of the curve is slightly above the gum reference. With increasing to 3 mm the slope becomes steeper. But for 6 mm initial length the reinforcement

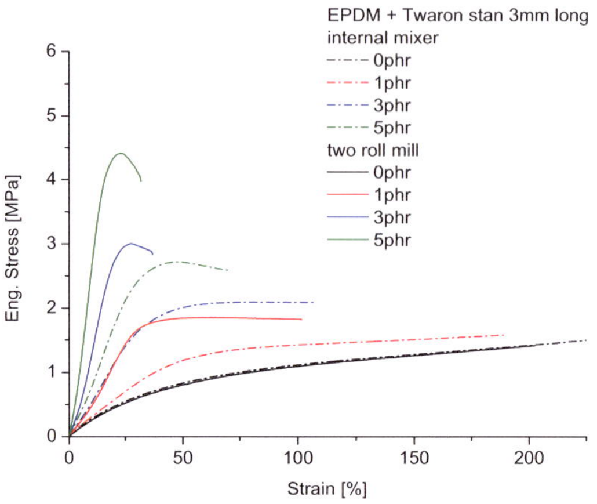

Figure 4.44: Influence of mixing device and fibre content on resulting stress-strain curves for Twaron fibres with standard finish in EPDM

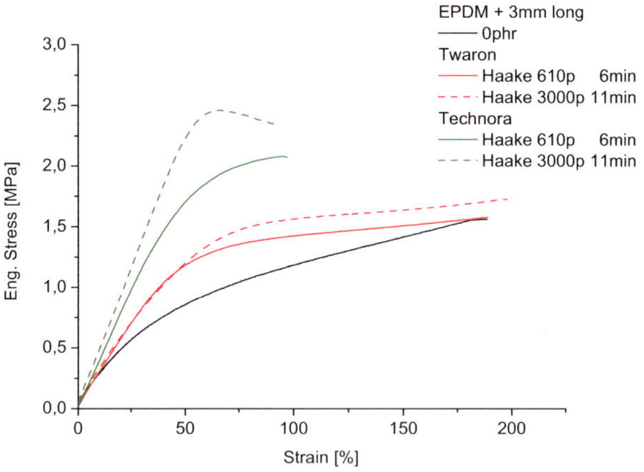

Figure 4.45: Influence of upscaling on resulting stress-strain curves for 1 phr Twaron and Technora with standard dip

decrease to a level between 1 and 3 mm, because the fibre length was shorten that much.

For Technora the initial length plays an important role since it is dominating the reinforcement. Technora, which was mostly unaffected by the processing showed a steeper initial slope of the stress-strain curve and significantly reduced elongation at break compared to Twaron. The high fibre length of Technora reinforced that much, that elongation at break was dropped under 100%.

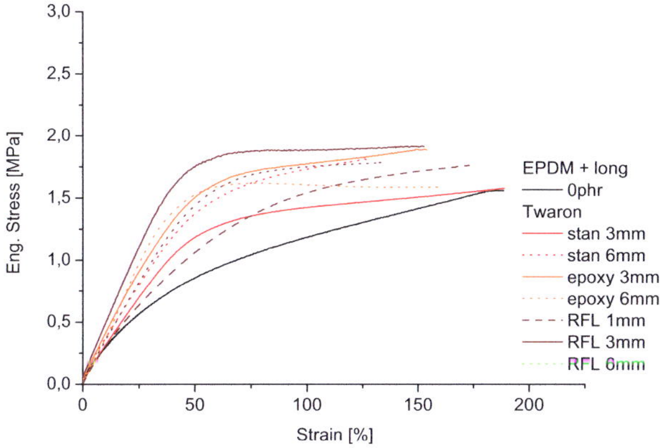

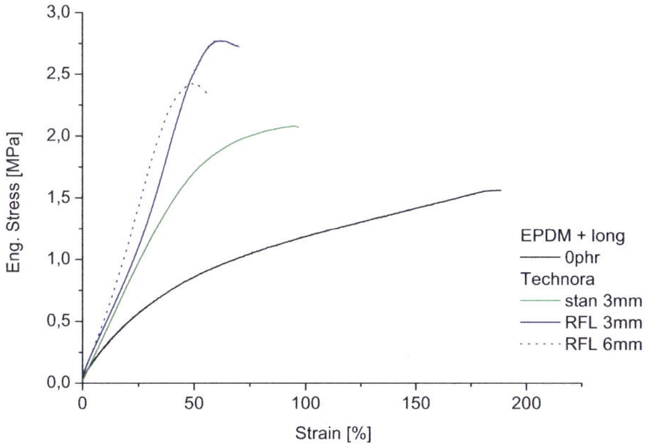

Figure 4.46: Influence of Twaron and Technora initial length on resulting stress-strain curves for 1 phr fibre content

Figure 4.47 shows the Young's modulus and elongation at break in fibre direction versus the weighted fibre length l_w after both processing steps with respect to initial length of the fibres. Untypically for the elastomers a Young's modulus had to be detected, since the reinforcement of the fibres was only significant for low elongations. This elastic modulus was determined as a secant modulus between 0 and 1% elongation.

The correlation for l_w to Young's modulus matched in case of Technora fibres, since the fibre breakage was not high. With increasing fibre length also the Young's modulus was shifting to higher levels. For Twaron, the samples assembled irrespective of the dip near to Young's modulus level of the 3 mm Technora specimen with standard dip. Exception were the standard dipped fibres and the one 1 mm RFL sample, which were lower and had also lower length and the Twaron RFL with 3 mm, which had the highest Young's modulus. This can be attributed to the additional effects of the agglomerates mentioned in the dispersion part.

The elongation at break in fibre direction decreased the higher the residual weighted fibre length was. Similar to the results of the Young's modulus this effect was much more pronounced for Technora fibres with reductions down to 60% strain at break. In case of Twaron the strain at break values were closer to that of the fibre free EPDM matrix. Due to the high scattering the specimen with 1 phr Twaron and RFL dip was even exceeding the value of 200%.

Finally, the influence of fibre type and matrix for 1 phr of fibres treated with standard or RFL dip mixed in EPDM or NR model systems by internal mixer can be seen in figure 4.48. The representative average stress-strain curves for the samples are displayed as engineering stress versus strain. The upper figure 4.48 shows the EPDM filled with 1 phr of Technora or Twaron with various treatments. As mentioned before Technora showed less fibre breakage and therefore reinforced better, which results in a steeper initial slope and dramatically reduced elongation at break. For both fibre types the RFL treatment reinforces most compared to the other dips.

NR filled with 1 phr of Technora or Twaron and various treatments is displayed in the lower part of figure 4.48. Compared to the EPDM this time the fibre breakage influence is observable clearly. Because of the high mixing times to disperse the fibre bundles completely and the mastication step, the remaining aspect ratios were relatively low. Even there was difference between the lengths of the fibre types as observable in figure 4.16, similar curves like the unfilled reference sample could be seen for the fibre filled specimen. Because of the significant fibre breakage also the differences between the fibre treatments are diminishing for Twaron. The correlation of the l_w value from the fibre length determination after both processing steps to the Young's modulus from tensile testing with respect to used matrix, fibre type and dip is shown in figure 4.49. For NR the significant increase of fibre breakage of Twaron and therefore lowered mechanical reinforcement compared to Technora was observable. In case of EPDM the unexpected differences in mechanical reinforcement were observed. Twaron fibres with RFL dip showed highest reinforcement values having lower fibre length as Technora fibres with standard finish and even less than Technora fibres with RFL dip. This can be explained by the increased anchoring effect of the agglomerates of Twaron fibres with RFL finish in the matrix (figure 4.7 (page 57)). In NR due to the higher mixing times or total strain applied the number and size of these agglomerates was negligible.

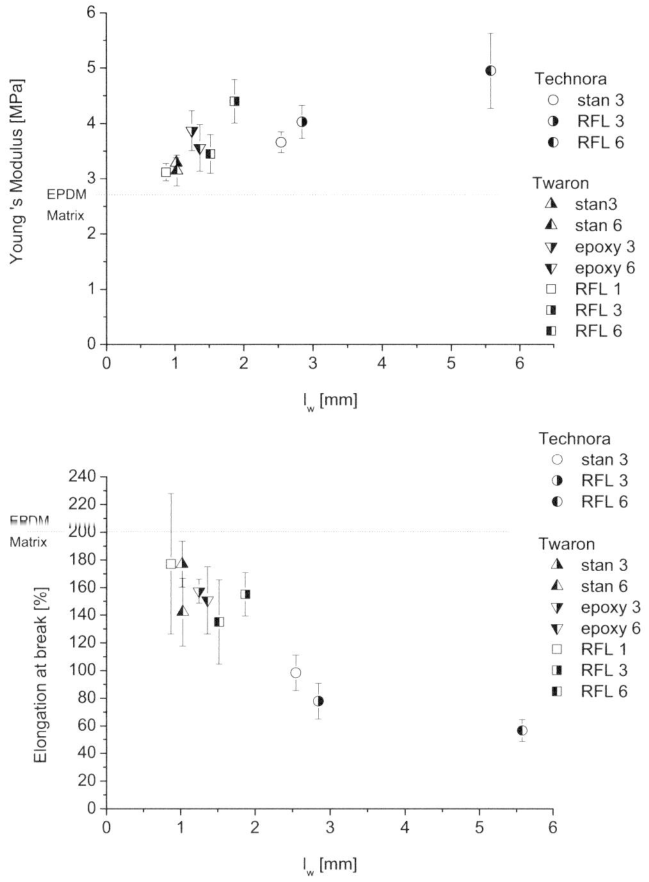

Figure 4.47: Young's modulus and elongation at break in fibre direction for 1 phr in dependence of l_w for different initial lengths of Twaron and Technora with different dips

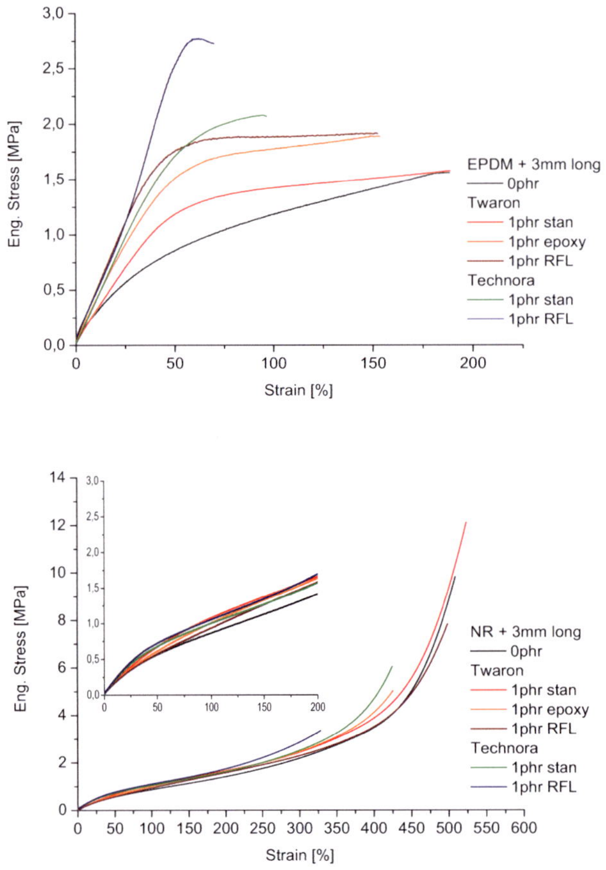

Figure 4.48: EPDM (top) and NR (bottom) filled with variations of fibre types and treatments

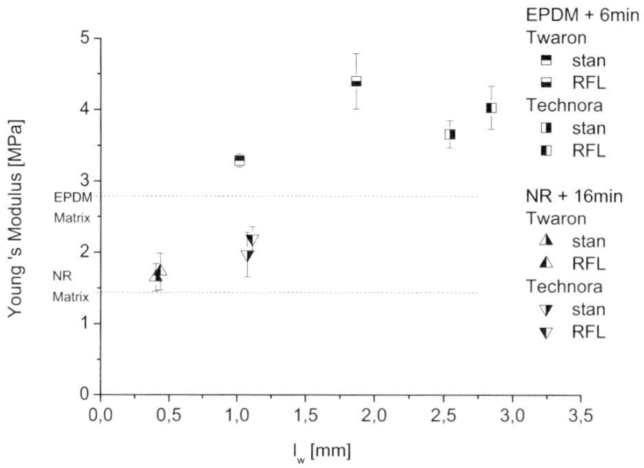

Figure 4.49: Young's modulus in fibre direction for 1 phr fibre content in dependence of l_w for Twaron and Technora with standard or RFL dip in NR or EPDM

4.7.2 EPDM and NR industrial systems

The dependence of the tensile curves for EPDM hose compounds on the Twaron fibre content and orientation can be seen in figure 4.50. The formulation without fibres is showing a significant kink which is observable for a strain of around 300%. This can be explained by the necking of the specimen which led to a virtual increase of stress values. However, similar to the model systems with increasing the fibre volume fraction the initial slope or Young's modulus is getting higher. In contrast to the model systems the strain at break is not significantly influenced by the fibre content. The transverse oriented samples show similar curves like the unfilled reference, while strain at break values decreased for higher fibre contents. The comparison of tensile curves between the EPDM model system and industrial

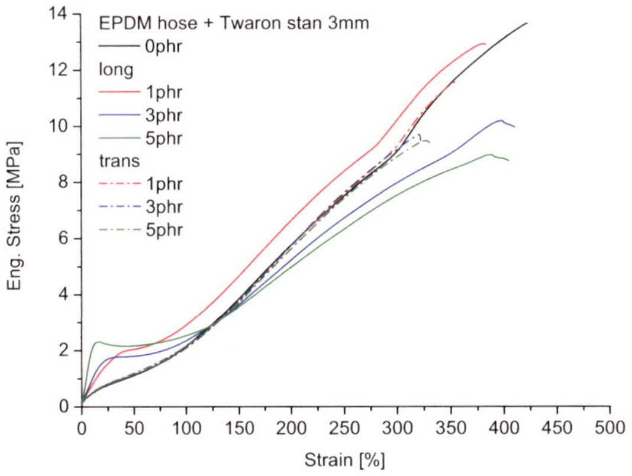

Figure 4.50: EPDM hose filled with variation of fibre content and orientation

system filled with variation of 1 to 5 phr standard fibres is shown in figure in 4.51. Although the fibre contents in phr for both systems were same the vol% differed remarkable like reported in table 3.2 (page 34). Since the industrial EPDM system was filled with carbon black and oil the difference between the stress-strain curves in terms of stress and strain values for both fibre free compounds was quite obvious. The elongation at break and the Young's modulus were for instance doubled from model to EPDM hose compounds in case of the fibre free specimen. However, the initial slopes of the curves showed similar shape for both systems. This can be attributed to similar amounts of fibre-matrix interface which is the product of remaining fibre length and volume content. Since the fibre volume content in the EPDM hose was roughly half that of the EPDM model system this means that fibres in the EPDM hose were definitely longer than those in the model compounds. An explanation for that is the significant lower viscosity of the the EPDM hose compared to the model system (table 4.4, page 51) which reduced the fibre breakage. Figure 4.52 shows the tensile curves for EPDM hose compounds filled with 5 phr of variation of fibre types and fibre dips. Like shown for the model systems Twaron RFL dipped fibres reinforced more than Twaron epoxy or standard dipped. Epoxy and RFL samples observed higher reinforcement in terms of initial slope by maintaining high elongation at break. Standard treated fibres obtained roughly half of the stress level at the end of the initial slope. Similar to the findings of the

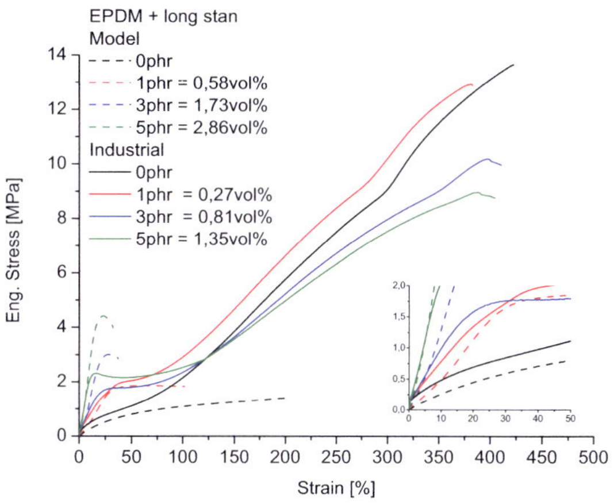

Figure 4.51: Comparison for EPDM model and EPDM hose compound filled with variation of fibre contents of Twaron fibres incorporated on two roll mill

model system Technora reinforced more than Twaron. Especially the RFL dip led to the highest initial slope with highest stress maximum about 5 MPa at the end of that slope.

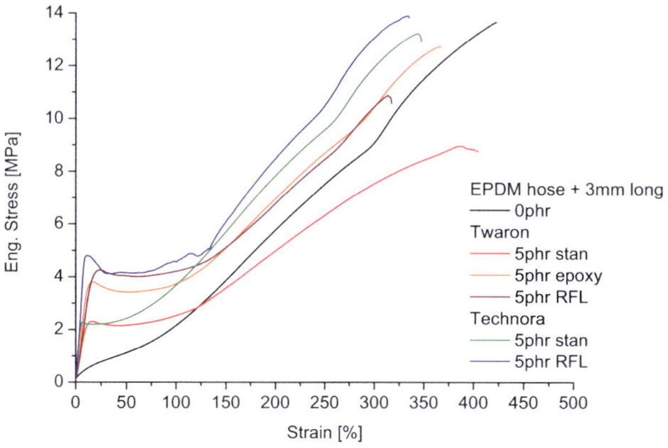

Figure 4.52: EPDM hose filled with variation of fibre type and treatment

Again a comparison between the tensile curves of the EPDM model system and industrial systems was made and is displayed in figure in 4.53. Although the fibre content differed again the shapes of the initial curves till the yield point were similar again, especially in case of Technora. For Twaron this can be explained by higher remaining length while the fibre volume content was just half of that of the model compounds. In case of Technora were the length should be similar, this fact is surprising, because the fibre volume content was lower.

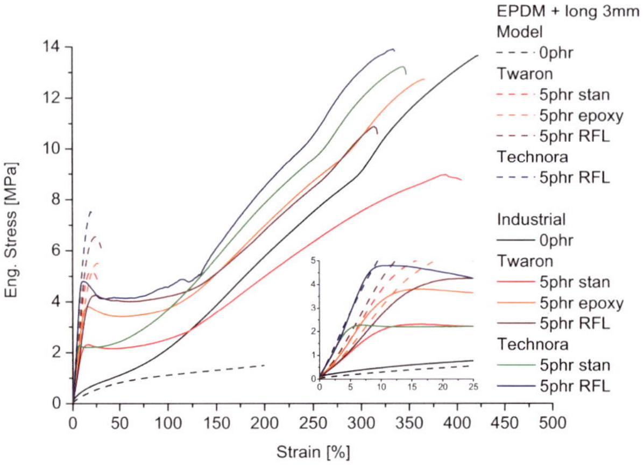

Figure 4.53: Comparison for EPDM model and EPDM hose compound filled with 5 phr of variation of fibre types and dips incorporated on two roll mill

The next figures shows the stress-strain curves for the NR tread compounds regarding fibre content, type, treatment and orientation. The reinforcement in terms of Young's modulus is low similar to the NR model compounds, but contrary to the expectation from the critical fibre length value in table 4.7 (page 76). This is caused mainly by the decrease of fibre length which is assumed to be similar as for the model compounds. However, the increase of fibre content leads to higher stress values in strain range between 20% and 150% for the longitudinal specimen. The stress values for the 5 phr transversal oriented specimen were slightly increased for a strain limit beneath 150% elongation compared to the reference and slightly decreased for strain values above this limit. This can be attributed to lowered orientation of the fibres in NR, like it was shown for the NR model system as well. The variation of fibre types and dips for NR tread composites can be seen in figure 4.55. Because of the findings for the model systems fibre treatments were just varied in case of Technora. As expected Technora fibres showed higher reinforcement in comparison to Twaron. Hereby the RFL dipped Technora fibres doubled the Young's modulus compared to Technora with standard treatment while keeping strain at break at similar level.

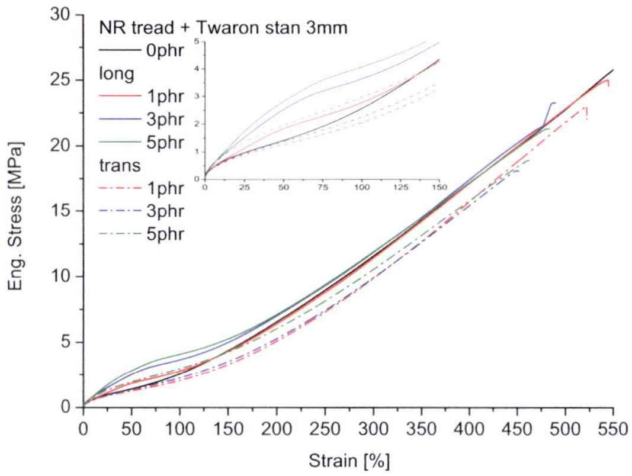

Figure 4.54: NR tread filled with variation of fibre content and orientation

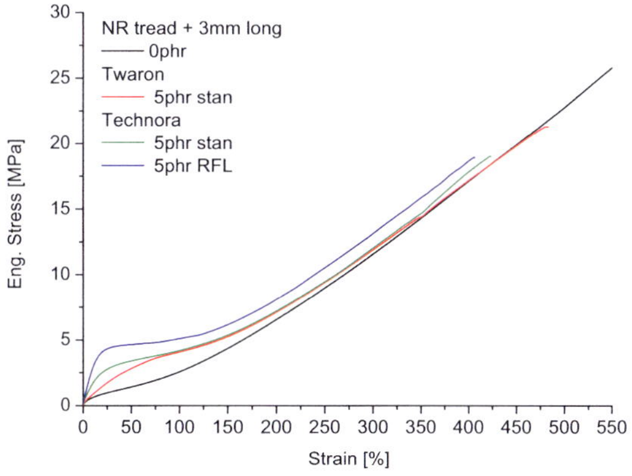

Figure 4.55: NR tread filled with variation of fibre types and treatments

Summary

The tensile testing behaviour of the transparent composites was investigated in dependence of matrix, fibre orientation, content, type and surface treatment. Also the influence of the process induced residual fibre length was studied. By increasing the fibre-matrix interface in fibre direction (longitudinal) through higher content or residual length the Young's modulus of the corresponding stress-strain curves got steeper and elongation at break decreased. Perpendicular to the main fibre direction (transversal) tensile curves followed that of the matrix on a slightly higher level, irrespective of the fibre content or length.

According to theories of short fibre reinforcement and the measured l_w the Halpin-Tsai values were calculated for the EPDM model compounds and displayed together with measured Young's modulus and Shore A hardness in figure 4.56. As expected Shore A hardness was rising from 52 to 60 with growing fibre content from 0,6 to 2,9 vol%. This increase was just depending on the fibre content and not on the fibre length or longitudinal or transversal orientation, since the hardness was measured with an indenter and the fibres were always laying in a plane perpendicular to the indentation direction. Also the measured Young's modulus was increasing with higher fibre contents. However, the measured values did not achieve the level of the Halpin-Tsai prediction in fibre direction (equation (2.5) page 7). Instead they were coming close to values predicted for a random orientation (equation (2.17)). One reason for that might be the insufficient fibre-matrix adhesion, which is assumed to be 100% in both models. Furthermore the strength of the fibres was reduced by the buckling and the reduction of fibre diameter through fibrillation similar as reported by Deteresa et al. [83]. Perpendicular to fibre direction the measured and calculated values fitted quite well. Due

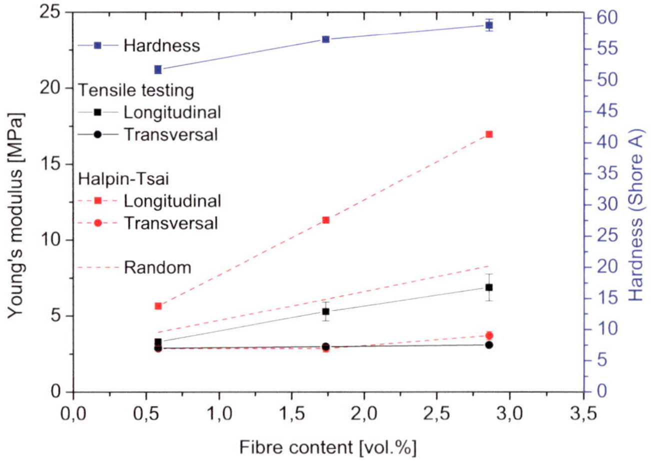

Figure 4.56: Young's modulus and Shore A hardness versus Twaron fibre content in vol% for EPDM model compounds mixed on internal mixer

to lower fibre breakage the fibres reinforced the EPDM matrix generally to a higher extent compared to those mixed in NR. In case of elastomer compounds containing carbon black

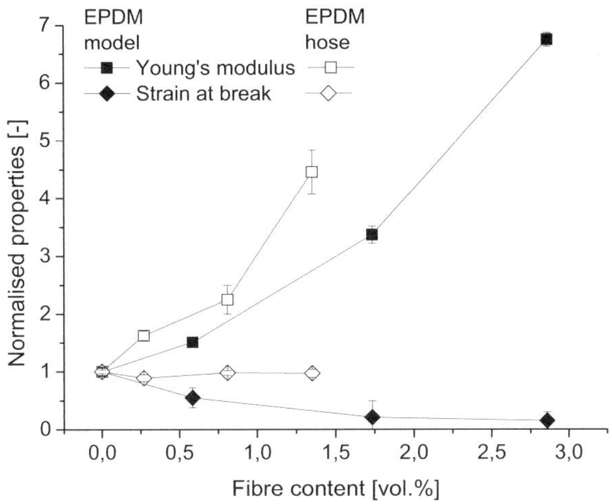

Figure 4.57: Normalised properties Young's modulus and strain at break in the composite and matrix versus fibre content for EPDM model and hose compound

the Young's modulus in fibre direction was also increasing with growing fibre content. The dependency of reinforcement on residual length influenced by fibre type and surface treatment was similar to the transparent compounds. In contrast to the findings of the model systems the elongation at break was not going down, mainly because of the presence of plasticizer oils. Due to this rise in stress levels by simultaneously constant strain at break, the area under the tensile curve was increasing, indicating a higher energy amount needed to tear the fibre reinforced compounds. Generally the reinforcement was more pronounced in EPDM most likely cause of lower fibre breakage compared to NR.

Figure 4.57 compares the normalised properties such as Young's modulus and strain at break for EPDM model and hose compounds in dependence of the fibre concentration. Normalised means in that way that ratios are displayed between the given properties of the composite in relation to the fibre free matrix. As reported strain at break stayed constant for the EPDM hose compound, while it decreased in case of the model compound. The level of Young's modulus was growing with higher fibre content, in case of the EPDM hose this trend was more pronounced compared to the model compound, which is indication of higher remaining fibre lengths and therefore reduced fibre breakage in the EPDM hose compounds. This can be attributed to the presence of the plasticizers in the industrial compounds which lowered the viscosity significantly.

The following 3D graphs correlate the measured value from fibre length distribution l_w and the values from CLSM α and o_f with the measured Young's modulus for the EPDM model systems from tensile testing (like shown in figure 3.11, page 45). Since there were three influence factors and one dependent value three images were generated. In the first picture the Young's modulus is correlated to the measured l_w and α. The red plane is representing the level of the predicted values of the Halpin-Tsai model (equation (2.16), page 8). The measured values did not reach the level of the Halpin-Tsai model, but followed the trend. As mentioned this is caused by the assumed 100% fibre-elastomer bonding and the reduction of fibre strength by buckling and fibrillation. With declining α and growing l_w the measured Young's modulus was getting higher. Still some datapoints did not fit the trend. That is why another two images below present the other possible correlations. The left one is showing the Young's modulus in dependence of l_w and o_f. As expected with increasing o_f and l_w also the modulus was rising. The image in the right shows finally the Young's modulus in dependence of o_f and α. The values increased with higher o_f and lower α values which was fitting quite well.

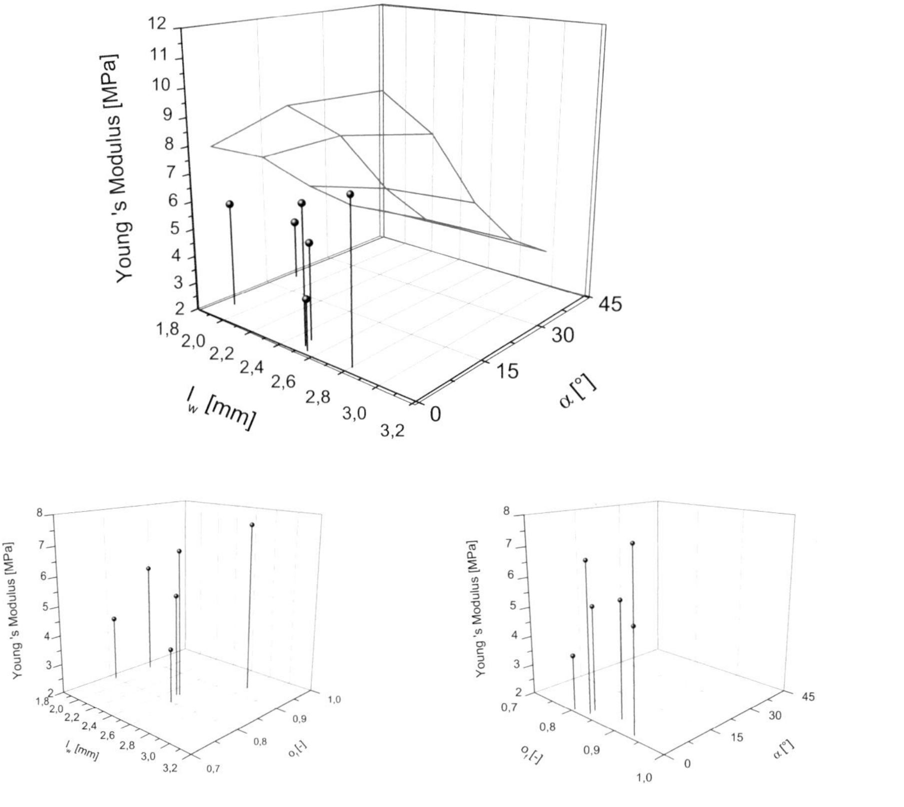

Figure 4.58: Correlation of the measured l_w, α and o_f values with measured Young's modulus

4.8 Dynamic mechanical analysis

4.8.1 EPDM and NR model system

The following figures show the results of the viscoelastic measurements in fibre direction for the investigated model systems. The dependence of Twaron fibre content is observable in figure 4.59. On the left side the storage modulus E' and loss modulus E" versus temperature are displayed. E' and E" levels were higher for growing fibre contents. The rise of E' was hereby higher than that of E". Therefore the peak of loss factor tan δ, which is the quotient from E" divided through E', was reduced by increasing Twaron fibre contents. The depen-

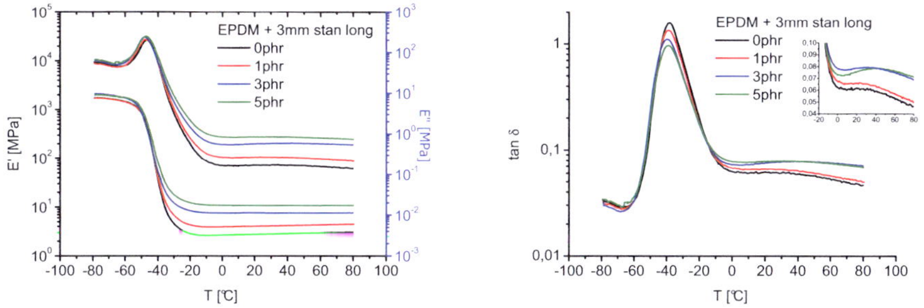

Figure 4.59: E' and E" versus T (left), tan δ versus T (right) for EPDM model with variation of Twaron fibre content

dence of mixing device for incorporation of the fibres is shown in figure 4.60. Similar as for the initial slope of the tensile curves, the two roll milled specimen showed higher reinforcement in terms of E', E" and reduction of tan δ peak. This behaviour was attributed to the fact that fibre length after final processing for fibres incorporated on internal mixer reached half of the values of those mixed in on two roll mill. To illustrate this figure 4.61 shows the

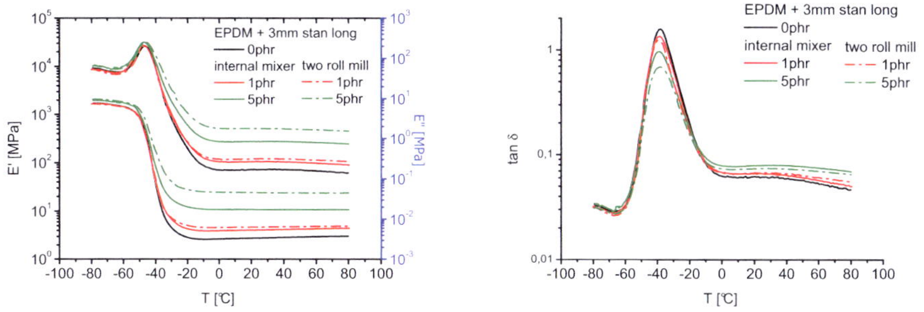

Figure 4.60: E' and E" versus T (left), tan δ versus T (right) for EPDM model with variation of Twaron fibre content and mixing device

ratio of tan δ peak of composite and matrix versus fibre content for both mixing devices. In

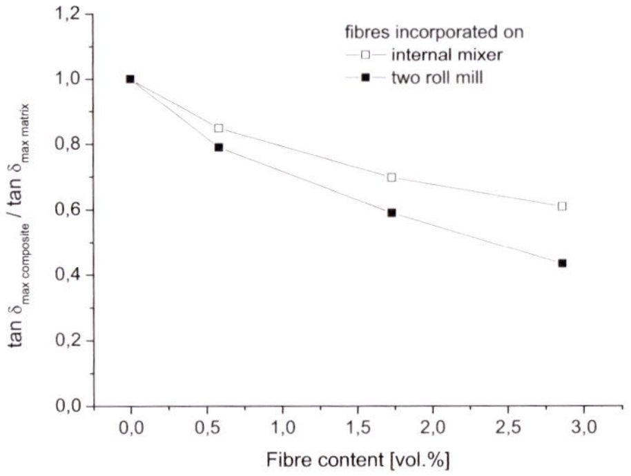

Figure 4.61: Ratio of tan δ peak in the composite and matrix versus Twaron fibre content for different mixing devices

case of fibres incorporated on two roll mill the descent of tan δ maximum was steeper with growing fibre content compared to those mixed on internal mixer.

In Figure 4.62 E', E" and tan δ are displayed versus temperature regarding the different fibre types and dips. Differences between the fibre types and dips were less pronounced. Twaron and Technora fibres dipped with RFL obtained slightly higher E' and E" level compared to standard or epoxy of the same fibre type. The maximum of mechanical loss factor tan δ around T_g was decreased for all used fibre types and dips to same level. Since

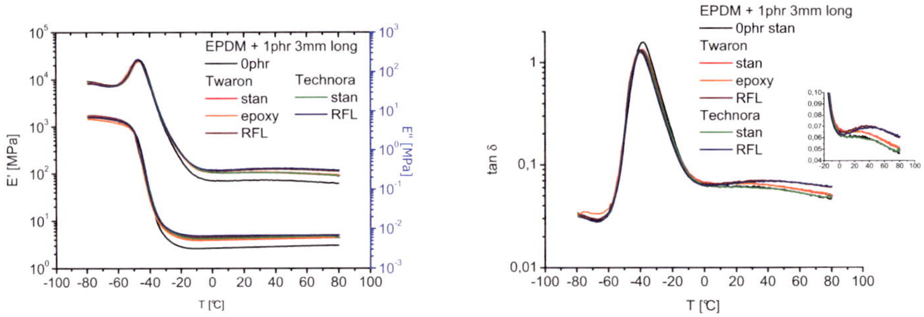

Figure 4.62: E' and E" versus T (left), tan δ versus T (right) for EPDM model of fibre types and dips

for NR model compounds due to dispersion problems no fibre content variation could be executed the influence of different used fibre types and dips on E' and E" and tan δ are displayed in figure 4.63. According to the findings for the fibre length Technora had the highest fibre length in the NR model compounds. Therefore also E' and E" levels were increased for the Technora compounds. This caused also that the corresponding tan δ peaks were reduced to a higher extent that the specimen filled with Twaron.

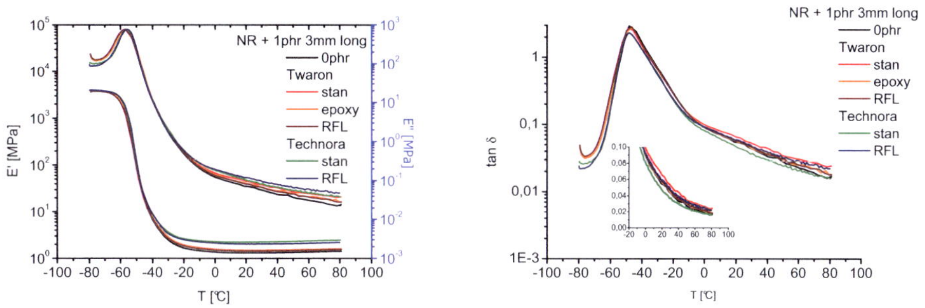

Figure 4.63: E' and E" versus T (left), tan δ versus T (right) for NR model of fibre types and dips

To summarise and compare the normalised ratio of $\tan\delta_{max}$ of the composite and the matrix for different investigated matrices and fibre types figure 4.64 gives an overview. In case of EPDM model system no significant differences between the loss of $\tan\delta_{max}$ for the different fibre types and dips were observable, while for NR the higher weighted fibre length of Technora led to higher decrease of the $\tan\delta$ peak. The drop of normalised ratio of $\tan\delta_{max}$ represented by the dotted line was slightly higher for NR compared to EPDM model compounds which is caused by the slightly higher fibre volume content in NR model compound (table 3.2, page 34).

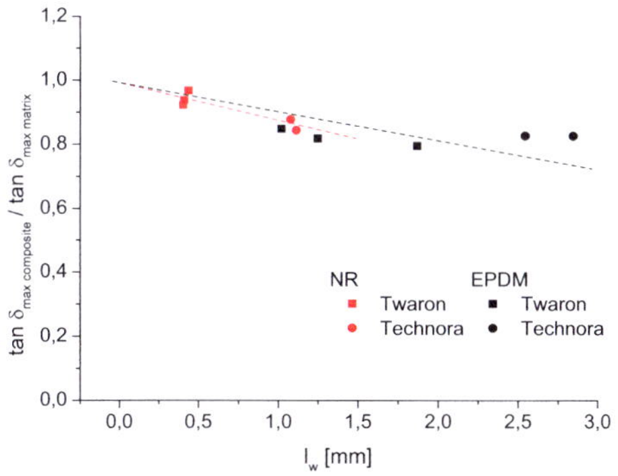

Figure 4.64: Ratio of tan δ peak in the composite and matrix versus weighted fibre length l_w for constant fibre content of 1 phr

In the next step ampsweeps were executed to determine the amplitude dependent viscoelastic behaviour of the short fibre reinforced elastomers. The following experiments show the storage modulus E' versus dynamic strain for the EPDM model compounds. The left of figure 4.65 shows the influence of orientation and Twaron fibre contents on E', while in the right the influence of the different mixing devices in case of longitudinal specimen for different Twaron fibre contents is displayed.

For the samples with fibre orientation perpendicular to testing direction the storage modulus E' was constant and slightly above the unreinforced matrix irrespective of the fibre content. For increasing fibre content in fibre direction the E' level of the samples was rising similar to the findings of the temperature dependent viscoelastic samples. With raising dynamic strain the E' values declined, which indicated the pulling out of fibres. A similar behaviour is well known as Payne effect [21] for carbon black filled elastomers. Also for elastomer composites filled with polyester short fibres [23] or carbon nano tubes [164] this effect was reported. In terms of mixing devices similar to the findings of tensile test the reinforcement in terms of E' and the decline with increasing dynamic strain was higher for specimen with fibres incorporated on two roll mill compared to that on internal mixer with same fibre content.

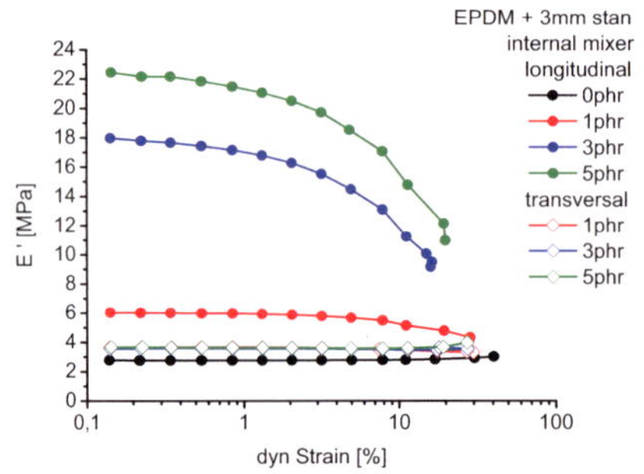

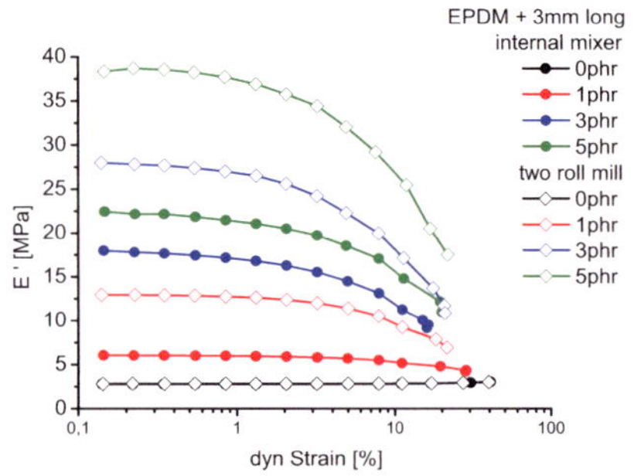

Figure 4.65: E' versus dynamic strain for EPDM model system: with variation of fibre content and orientation (left) and variation of fibre content and mixing device (right)

4.8.2 EPDM and NR industrial systems

The temperature dependent viscoelastic properties of the short fibre reinforced carbon black filled compounds in fibre direction are shown in the next figures. On the upper left side of figure 4.66 storage modulus E' and loss modulus E" are displayed versus temperature for EPDM hose compound with variation of Twaron fibre contents. With increasing the Twaron fibre content above 1 phr the obtained E' and E"levels were rising. Similar to the EPDM model systems the peak of loss factor tan δ was reduced by increasing Twaron fibre content, because the rise of E' was dominating the one of E" (Figure 4.66 upper right). Contrary to the EPDM model systems the level of the tan δ curves for the fibre filled compounds were laying under that of the gum compound above a temperature of 0°C.

In the lower part of figure 4.66 storage modulus E', loss modulus E" and mechanical loss factor tan δ versus temperature for EPDM hose compounds are displayed regarding the used fibre types and treatments for a constant fibre content of 5 phr. This higher fibre content was chosen, since the fibre volume content was lower in the industrial systems and differences to the matrix were just observable above 1 phr. E' and E" levels were rising. The order was hereby related to the order of the fibre length after processing found for the model compounds. The reduction of the tan δ maxima around T_g was also in that order. Especially in the case of Technora the peak was broadening and shifted to lower temperatures.

In the following the temperature dependence of the NR tread samples was tested with

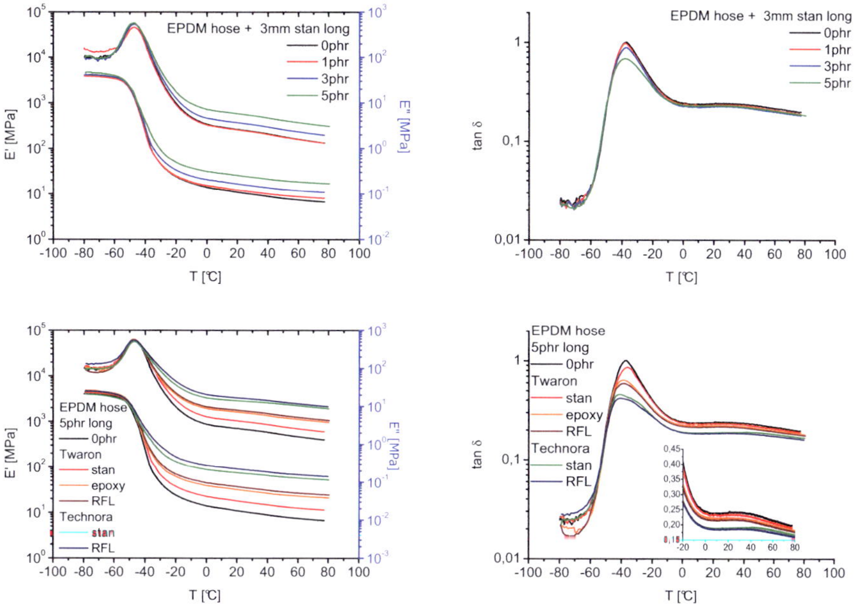

Figure 4.66: E' and E" versus T (left), tan δ versus T (right) for EPDM hose compound with variation of fibre contents (top), fibre types and treatments for 5 phr fibre content (bottom)

respect to the fibre content and fibre types and dips. E', E" and tan δ versus temperature regarding the Twaron fibre content are shown in upper part of figure 4.67. With increasing fibre contents also storage and loss modulus levels were rising and therefore the level of tan δ curves were declining with growing fibre content.

In the lower part of figure 4.67 the viscoelastic properties as function of temperature are shown in dependence of the used fibre types and treatments for a constant fibre content of 5 phr. The difference between the used fibre types and fibre treatments in terms of E', E" and tan δ were moderate. The tan δ decrease was slightly higher for the Technora samples compared to the Twaron sample. The curves for Technora samples showed a to a lower temperature shifted maxima and a shoulder around -25°C.

To summarise and compare the measured $\tan\delta_{max}$ values and the calculated reductions in % with regard to matrix, fibre type and treatment can be seen in table 4.12. For both investigated systems with growing Twaron fibre content the decrease of $\tan\delta_{max}$ was as expected increasing. Technora fibres gave higher drop of $\tan\delta_{max}$, because of their higher residual lengths in the composites. The RFL dip enhanced this effect. For the NR tread compound also the loss of $\tan\delta$ at areas of 0 or 60°C corresponding to the predictors of wet grip and rolling resistance for tyres [165] were displayed together with their calculated loss. With increasing Twaron fibre content also the rolling resistance was slightly going down. But at the same time the ability for wet grip was sacrificed as observable for 5 phr. For Technora with standard finish both values were surprisingly higher as the reference. However, in case

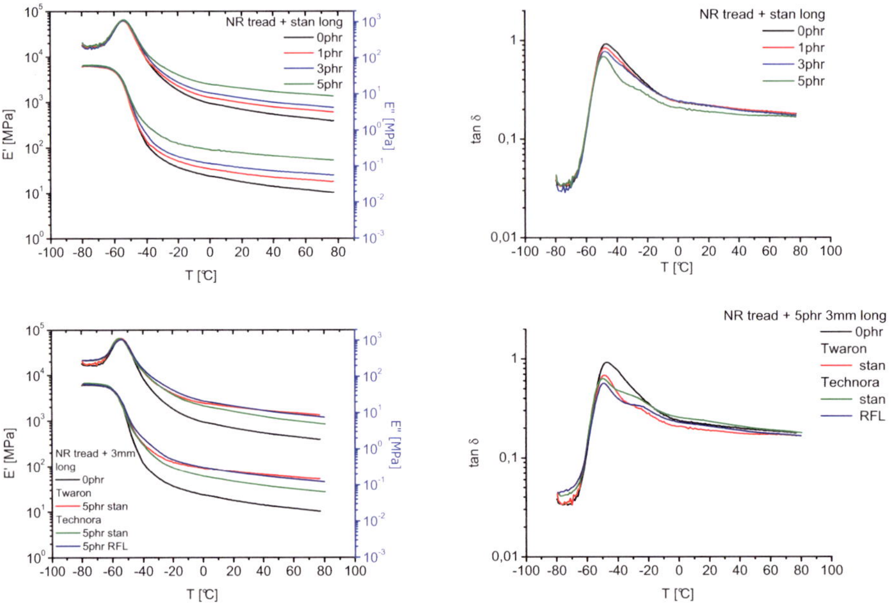

Figure 4.67: E' and E" versus T (left), tan δ versus T (right) for NR tread compound with variation of Twaron fibre content (top), of fibre types and treatments for 5 phr fibre content (bottom)

of RFL the reduction was observable again, but on much lower level compared to specimen with 5 phr Twaron. The reason was the different shape of the curves for the temperature dependent viscoelastic values displayed in figure 4.67.

	tan δ [max]	Loss [%]		tan δ [max]	Loss [%]	tan δ [60°C]	Loss [%]	tan δ [0°C]	Loss [%]
EPDM hose	1,011		NR tread	0,921		0,187		0,238	
Twaron			*Twaron*						
1 phr stan	0,979	3	1 phr stan	0,840	9	0,187	0	0,238	0
3 phr stan	0,885	12	3 phr stan	0,768	17	0,181	3	0,238	0
5 phr stan	0,857	15	5 phr stan	0,681	26	0,172	8	0,206	13
Technora			*Technora*						
5 phr stan	0,459	55	5 phr stan	0,630	32	0,191	-2	0,256	-7
5 phr RFL	0,420	59	5 phr RFL	0,564	39	0,175	6	0,228	4

Table 4.12: tan δ_{max}, tan δ at 0 or 60°C and calculated loss

Like for the model systems ampsweeps were executed also to determine the amplitude dependent viscoelastic behaviour of the short fibre reinforced industrial systems. Figure 4.68 shows the influence of orientation and fibre content on storage modulus of the compounds. For the samples with fibre orientation perpendicular to testing direction the storage modulus E' was slightly above the unreinforced matrix. For samples tested in fibre direction with growing fibre content the E' levels were shifted to higher levels. With increasing dynamic strain the E' values dropped as expected. Also for the NR tread samples the amplitude dependent behaviour was investigated. Some of the transversal specimen reached the level of 1 phr longitudinal and the 5 phr transversal sample came close to the level of 5 phr longitudinal, which is indication of insufficient orientation.

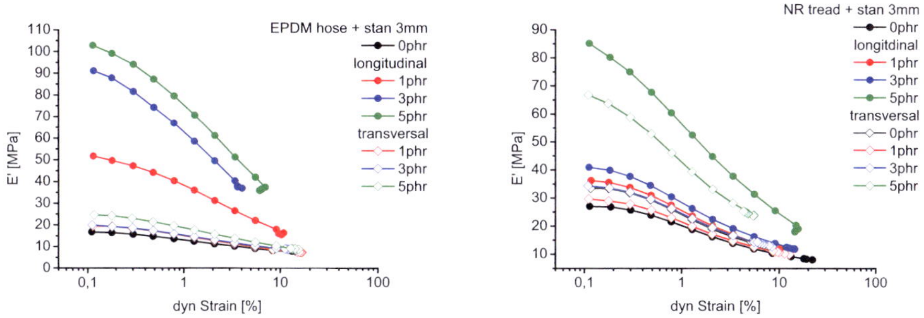

Figure 4.68: Storage modulus versus amplitude for EPDM hose with variation of fibre content and orientation

Summary

The dynamic viscoelastic properties in dependence of temperature were investigated for the model systems regarding the matrix and fibre parameter as content, type and surface treatment as well as the influence of mixing device and residual fibre length. With increasing the fibre-matrix interface by a higher fibre content or residual length the storage modulus and loss modulus levels were rising. Therefore the resulting peak of loss factor tan δ at the glass transition temperature T_g was going down. With growing dynamic strain amplitude the fibres were pulled out of the matrix and the storage modulus was decreasing.

For the carbon black filled elastomer compounds similar trends could be observed regarding the fibre content dependence of storage modulus, loss modulus and loss factor. Also findings for the used fibre types and surface treatments were similar to those of the models systems indicating that the trends of fibre breakage behaviour were similar.

4.9 Heat build-up

Flexometer tests were executed to industrial carbon black filled compounds according to DIN 53533 [154], which assumes tests under the condition of constant strain. This condition is valid when thinking of applications like seals. By adding fibres to these compounds the hardness was increased like reported in figure 4.56 (page 98). This hardness increase required a higher force to compress the fibre reinforced specimen in comparison to the fibre free reference, which led to an increase of heat build-up and compression set. This effect can be seen in figure 4.69. The temperature increase inside the samples during the test, the compression set after testing and their initial Shore A hardness values are displayed. For

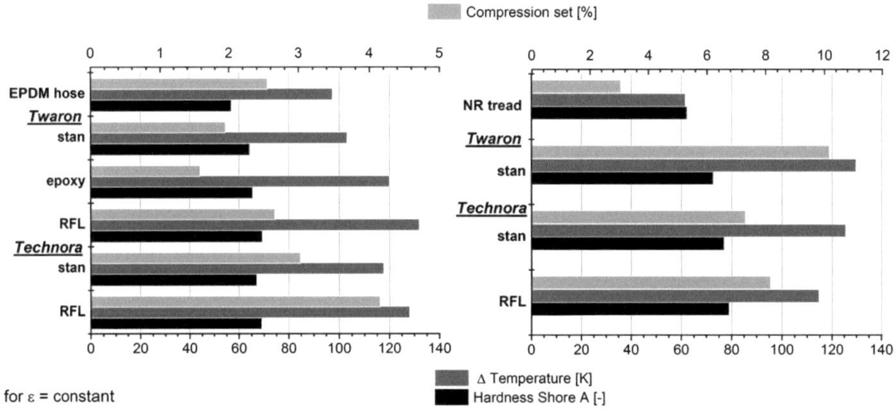

Figure 4.69: Hardness, compression set and temperature increase during heat build-up of EPDM hose and NR tread specimen for constant strain

the EPDM hose compound the Shore A hardness increased from 57 for the reference to 69 for compounds filled with 5 phr Twaron or Technora dipped with RFL. With growing hardness also the temperature increase jumped from 98 K for the reference to around 130 K for the specimen with highest hardness. The permanent compression set after testing nearly doubled from 2.55% to over 4%. The hardness rise due to fibre addition was higher for the NR compounds compared to EPDM. While the NR reference specimen had a Shore A hardness of 62, those filled with 5 phr Technora obtained values close to 80. Although the sample with 5 phr Twaron had just a hardness of 73, it displayed the highest temperature increase and compression set, with 130 K and 10% respectively.

To investigate also the situation for a tyre or a timing belt, tests were also carried out under the condition of constant stress. This led to a decrease of heat build-up level and compression set. The harder fibre reinforced samples achieved a lowered elongation for a constant stress, which caused a lower heat build-up in these compounds. Figure 4.70 shows that there is minor drop of temperature increase and compression set for the fibre filled compounds. In case of EPDM the temperature increase of the fibre filled samples was always lower than that of the reference. The compression set, however, was on the same level or slightly above it. For NR the same trend was observed. The compression set and temperature increase stayed below the level of the reference, except for the sample with filled with 5 phr Technora with standard dip. Beside this deviation it could be shown that the fibres have no negative effect on fatigue and thermal stability in NR treads for the condition of constant pressure.

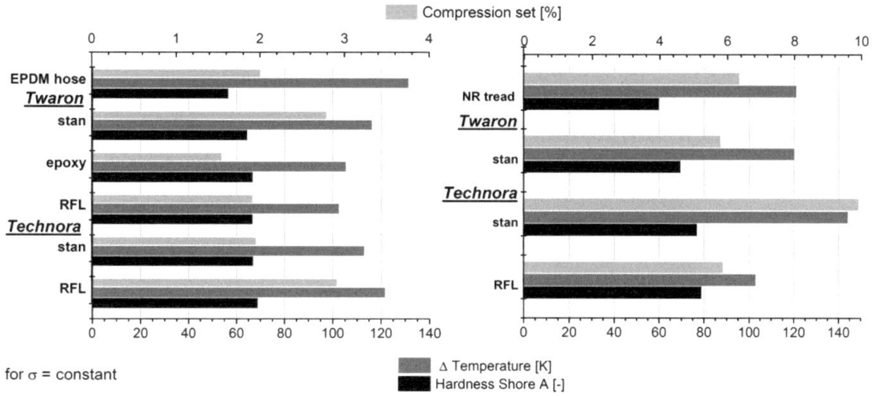

Figure 4.70: Hardness, compression set and temperature increase during heat build-up of EPDM hose and NR tread samples for constant stress

4.10 Tear fatigue analyser

To predict the service lifetime of the vulcanisates tear fatigue analysis was carried out. Due to the significant higher area under the stress-strain curve predicting higher tear energy a fibre reinforced EPDM hose compound was chosen.

Specimen of the EPDM hose compound without and with 5 phr Twaron standard fibres in fibre direction and perpendicular to the crack propagation direction were tested for five different amplitudes. Tearing energy (T) and crack propagation rate (da/dn) were recorded.

Figure 4.71 shows data points obtained at the different amplitudes (amplitude values are displayed near the data points) and the linear fits in a graph of logarithmic tearing energy versus logarithmic crack propagation rate.

For the sample with fibres oriented in crack direction, the crack propagation rate was slightly increased in comparison to the fibre free reference. The specimen with fibres oriented perpendicular to the crack direction obtained a propagation crack rate which was decreasing for amplitudes higher than 14,51%. This trend was confirmed by the lower slope of the linear fit and was caused by the disorientation of the crack propagation front along the perpendicular orientated fibres and therefore increased crack length.

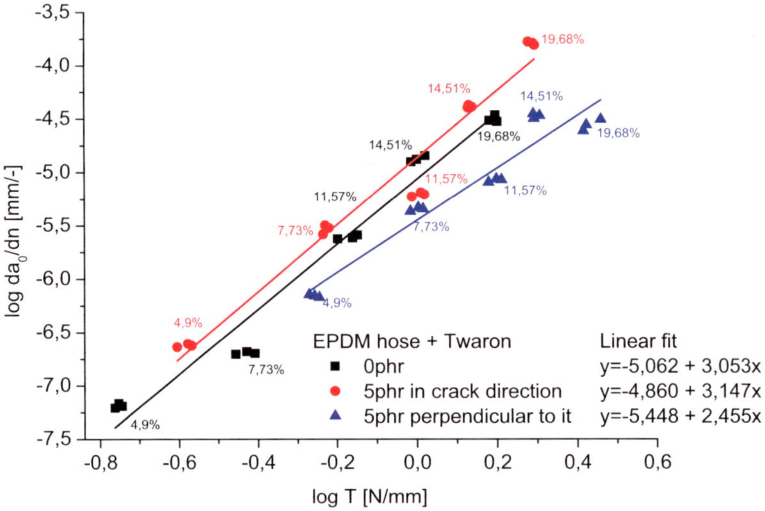

Figure 4.71: Logarithmic tearing energy versus logarithmic crack propagation rate for EPDM hose compound with 5 phr different oriented Twaron fibres

Chapter 5

Conclusions

5.1 Summary and Conclusions

The aim of the present work was to investigate and correlate the processing and material influences on the morphology and resulting properties of short aramid fibre elastomer composites.

Aramid fibres were chosen because of their resistance against occurring high shear stresses during mixing of elastomers. Different fibre treatments were applied to improve the fibre-elastomer interaction. Model gum stocks were studied to understand the morphology-property relationship and correlate the found trends to carbon black filled elastomer systems, for which morphology detection was hampered. Each influencing factor, whether fibre or processing related, was monitored through the whole morphology-property chain, regarding compounding process and fibre dispersion, length, diameter and orientation distribution, fibre-elastomer interaction, curing, tensile testing, dynamic mechanical and fatigue behaviour. Procedures for detecting the fibre length distribution and orientation were created and applied.

In general the fibre dispersion was dependent on the specific energy input. For higher fibre contents or lengths also higher specific energy input were needed to achieve sufficient state of dispersion. Upscaling the dispersion quality from smaller to bigger sized internal mixers was possible, but the needed specific energy input was enlarged. No significant influence of viscosity could be detected for EPDM model compounds. The utilised fibres dispersed earlier to filament level in EPDM compared to NR, because its elasticity was lower.

Twaron fibres showed a specific energy input dependent shortening behaviour, while Technora fibres were relatively unaffected. Therefore the initial fibre length for Technora had a direct impact on the reinforcement and in case of Twaron there was no significant influence. Differences in Twaron fibre length after processing step one, the fibre incorporation, were superimposed by the second processing step, the incorporation of curing additives and orienting of the fibres on two roll mill. However, fibres processed only on two roll mill showed doubled residual fibre length compared to those mixed on internal mixer and two roll mill. The aspect ratio of the fibres was not only changed by lowering the filament length, also a diameter reduction in form of fibrills was observed.

Due to the higher specific energy input needed for dispersion in NR and the specific mastication behaviour, Twaron shortened to a higher degree and therefore anisotropy level and reinforcement level were relatively low compared to EPDM compounds.

Mooney viscosities were not affected to significant extent by adding the investigated low amounts of fibres. Similar were the findings of the curing behaviour. Adding fibres increased the maximum torque levels and lowered the t_{90} optimum curing times. This was caused by faster heat transfer into the compound through the fibre-elastomer interface and higher internal friction of the compounds.

The differences regarding morphology between the oily standard finish and the epoxy dip were not significant. For Twaron the applied RFL dip was leading to agglomerates or lumps consisting of several filaments. Because these multiple filaments could better resist against buckling, the residual fibre lengths and the final reinforcement effect was higher compared to the other surface treatments. Trials to improve the fibre-elastomer interaction by irradiation for postcrafting of the Twaron fibre surfaces failed due to the higher fibre breakage level during mixing. Generally a direct proof for increased fibre-elastomer adhesion could not be found since the fibre-matrix interface failed above a certain strain level in form of fibre pull-out.

The mentioned adhesion behaviour led to a characteristic yield point in the tensile tests where the fibre-elastomer interface failed and the load was carried by the elastomer matrix alone. The yield point was visible in the region of small strain under 50% for composites tested in fibre direction. The higher the fibre-elastomer interface, the product of residual fibre length and -content, the higher the yield point and the lower the strain at break of the model composites. The Young's modulus of the composites was correlating with orientation and the fibre-elastomer interface. For fibres oriented perpendicular to the direction of tension the yield point was not found. These specimens showed stress levels slightly above the matrix while strain at break was relatively unaffected. Similar to that the orientation and fibre-elastomer interface influenced the reduction of the loss factor peak of the dynamic mechanical measurements in dependence of temperature. The degree for the drop of storage modulus level with increasing dynamic strain was also affected.

Orientation of the fibres in dependence of the position in the moulds was successfully investigated by optical methods for transparent model systems. To quantify the orientation two predictors were introduced: an orientation factor and a deviation angle to the direction of applied tension. Both indicators were used together with the measured weighted fibre length to predict the mechanical behaviour in terms of Young's modulus. Measured results followed the trend for the Halpin-Tsai predicted values, but were in fibre direction lower as calculated. The reasons were the weakening of the fibre due to buckling and fibrillation and the insufficient fibre-elastomer adhesion.

Since carbon black, plasticizers and other additives were used in these industrial recipes the relative fibre volume content was lower and the morphology detection was hampered. However, the found trends for the residual fibre lengths and dispersion levels of the model and industrial systems were similar, which was validated partly by SEM analysis. Indirectly also the similar behaviour of tensile and DMA results in dependence of fibre content, type and dip confirmed that results could be transferred from model to industrial compounds. Due to the normalised results in dependence of fibre volume content and the lower viscosities induced by the plasticizers, it was assumed that the fibre breakage in the in carbon black filled systems was lower and the residual fibre lengths higher. Generally the Shore A hardness was increasing with growing fibre content for all compounds. The fatigue behaviour of the carbon black filled vulcanisates was improved by adding fibres. In case of heat build-up also the measurement conditions (constant strain or stress) had a significant influence. The results of the tear fatigue analysis showed a crack deflection, if the fibres were oriented perpendicular to crack propagation direction.

Finally, it was possible to study the influence of different processing and fibre parameter on the morphology of transparent short aramid fibre reinforced elastomers by new established optical characterisation methods. Contrary to other works the different contributions of the morphology as fibre dispersion, aspect ratio, orientation and fibre-elastomer interaction on the final composite properties could be successfully distinguished and linked to each other. In the end the found relationships helped to understand the influence factors on the properties of industrial relevant carbon black filled elastomer compounds. As mentioned in the beginning this thesis is part of Dutch Polymer Institute (DPI) project 664 SUSTFIBER "Sustainable elastomers and thermoplastics by short-cut fiber reinforcement" and is connected with two other theses. The results of this work might help to evaluate the findings of these theses dealing with ways to improve the fibre-elastomer adhesion and the tribological behaviour of carbon black filled elastomer compounds.

5.2 Outlook

The presented results can give only a limited view into the topic. Several interesting routes and open questions appeared during the work on this thesis.

Masterbatch methods or usage of compatibiliser like for example graft-copoly (ethylene/maleic anhydride) as mentioned in the chapter state of the art were not studied, because of possible negative effects on the optical characterisation methods for determining the fibre morphology. However, the degree of improvement on fibre dispersion, length and ultimate properties would be an interesting target to investigate.

Since no clear proof for the improved fibre-elastomer adhesion of the different applied fibre surface treatments could be found, it would be useful to study the influence of aramid pulp on the morphology and especially the anisotropy of the composite properties. The established characterisation methods for studying the morphology of highly fibrillated fibres in transparent compounds would be feasible to investigate the dispersion and breakage behaviour of aramid pulp on micron scale.

To study the fibre orientation further it would be also feasible to apply the special multiple compression premoulding technique reported by Clarke and Harris [116] and investigate the degree of orientation improvement by CLSM and the introduced orientation predictors deviation angle α and orientation factor o_f.

It maybe also be helpful to verify the fibre morphology in the carbon black filled composites by a direct method as it was shown for one example with x-ray computed tomography in figure 3.10 (page 43). However, the high measuring time and reconstruction effort has to be considered to evaluate the cost-benefit relationship.

Finally, extended tear fatigue analysis would also be worth to investigate new combinations of different fibres orientations, aspect ratios and matrices to characterise their fatigue behaviour.

Chapter 6

Appendices

6.1 Calculation of phr, wt% and vol%

The fibre fraction of the composite in parts per hundred parts rubber one can get by dividing the fibre mass by the mass of the rubber used in the composite.

$$phr = \frac{m_{fibre}}{m_{rubber}} \times 100 \tag{6.1}$$

The fibre wt% in the composite is given by the fraction of fibre mass divided by the total mass of the composite.

$$wt\% = \frac{m_{fibre}}{m_{total}} \times 100 \tag{6.2}$$

The fibre vol% in the composite can be calculated to according to.

$$vol\% = \frac{V_{fibre}}{V_{total}} \times 100 \tag{6.3}$$

With the relation that a volume of any material is dependent of its mass and density,

$$V = \frac{m}{\rho} \tag{6.4}$$

we can combine the two equations to get the fibre vol%.

$$vol\% = \frac{m_{fibre}/\rho_{fibre}}{m_{total}/\rho_{total}} \times 100 \tag{6.5}$$

6.2 Theoretical prediction of composite modulus and critical fibre length

To calculate the composite modulus in dependence of fibre content, length and orientation the theories of Halpin-Tsai model and lamina theory have been used. Also critical length was calculated using the values from table 6.1 showing the fibre and matrix input parameters used. The results for one example can be observed in table 6.2.

Fibre			Source	*Matrix*			Source
ν_f	0,35		[159]	ν_m	0,5		literature
d_f	12	μm	SEM				
ϕ_f	variable		table 3.2				
E_{Twaron}	67000	MPa	[159]	$E_{EPDMmodel}$	2,83	MPa	tensile test
G_{Twaron}	1500	MPa	[166]	$G_{EPDMmodel}$	0,94	MPa	eq. 2.10
$E_{Technora}$	71000	MPa	[159]	$E_{NRmodel}$	1,48	MPa	tensile test
$G_{Technora}$	1500	MPa	[166]	$G_{NRmodel}$	0,49	MPa	eq. 2.10
				$E_{EPDMhose}$	4,39	MPa	tensile test
				$G_{EPDMhose}$	1,46	MPa	eq. 2.10
				$E_{NRtread}$	8,05	MPa	tensile test
				$G_{NRtread}$	2,68	MPa	eq. 2.10

Table 6.1: Input parameter

Result			Source
E_L	5,60	MPa	eq. 2.5
E_T	2,88	MPa	eq. 2.6
η_L	0,99		eq. 2.11
η_T	1,00		eq. 2.12
η_G	1,00		eq. 2.13
ν_{LT}	0,5		eq. 2.7
G_{LT}	0,95		eq. 2.9
E_1	4,73	MPa	eq. 2.16

Table 6.2: Results for a given a EPDM model compound filled with 1 phr (0,58 vol%) Twaron fibres with a l_w of 1 mm and orientation α of 15 °

Symbols and Abbreviations

Latin Symbols

A	Area of the rotors
a_0	Initial crack length
D	Rotor diameter
d	Fibre diameter
da/dn	Amount of crack propogation per cycle
e	Specific energy input
E'	Storage modulus
E''	Loss modulus
E_f	Elastic modulus of the fibre
E_L	Longitudinal composite modulus
E_m	Matrix modulus
E_T	Transversal composite modulus
E_{random}	Composite modulus for a random orientation
G_m	Shear modulus of the matrix
G_{random}	Composite shear modulus for a random orientation
G	Shear modulus
h	Rotor tip clearance
L	Rotor length
l	Fibre length
l_c	Critical fibre length
l_n	Number average of fibre length
l_w	Weighted average of fibre length
M	Torque
m	Mass
M_f	Mooney viscosity of fibre filled compounds
M_g	Mooney viscosity gum compound
M_W	Molecular weight
N	Rotor speed
$N_{bundles}$	Number of undispersed fibre bundles
N_{RS}	Dimensionless dispersion number
o_f	Orientation factor
r_f	Fibre aspect ratio defined as l/d
r	Vector between shifted images
S'	Torque of the curing measurement
t	Mixing time
T_g	Glass transition temperature

t_{90} .... Optimum curing time
t_{mix} .... Fibre mixing time
T .... Tearing energy
V .... Volume
V_0 .... Net chamber volume
V_S .... Shear volume
dW_E .... Amount of complete elastic energy
$vol\%$ .... Volume percent
W_a .... Work of adhesion
W_E .... Elastic energy density
$wt\%$ .... Weight percent

Greek Symbols
α .... Angle to the axis of applied stress
$\dot{\gamma}$ .... Shear rate
Γ .... Total strain
γ^D .... Disperse part of surface energy
γ^P .... Polar part of surface energy
γ_l .... Surface energy of the liquid
γ_{sl} .... Interfacial energy between solid and liquid
γ_s .... Surface energy of the solid
θ .... Contact angle
σ .... Stress
ϵ .... Strain
ΔG_i .... Free energy of immersion
ν_f .... Poisson's ratios of the fibre
ν_m .... Poisson's ratios of the matrix
ϕ_f .... Fibre volume fraction
ρ .... Density
σ .... Stress
σ_f .... Fibre tensile stress
τ_g .... Fibre-matrix interface strength
$tan\,\delta$ .... Loss factor defined as E''/E'

Abbreviations
AFM .... Atomic force microscopy
CLSM .... Confocal laser scanning microscopy
CR .... Polychloroprene rubber
DCP .... Dicumylperoxide
DMA .... Dynamic mechanical analysis
DOF .... Degree of fill
DPI .... Dutch Polymer Institute
DSA .... Double strain amplitude
ENB .... Ethylidene Norbornene
ENR .... Epoxidized Natural Rubber
EOC .... Ethylene octene copolymer
EPDM .... Ethylene-propylene-diene rubber

Epoxy .... Epoxy fibre dip
EPR .... Electron Paramagnetic Resonance Spectroscopy
ETE .... Departement of Elastomer Technology and Enginieering
EU .... European Union
FLD .... Fibre length distribution
FOD .... Fibre orientation distribution
IPF .... Leibniz-Institut für Polymerforschung Dresden e.V.
IR .... Isoprene rubber
LCP .... Liquid crystal polymers
$ML(1+4)100$ Mooney viscosity with large rotor at 100°C
MPIA .... M-aramid poly(m-phenylene isophtalamide)
MU .... Mooney units
NBR .... Acrylonitrile-co-butadiene rubber
NR .... Natural rubber
OTR .... Departement of Surface Technology and Tribology
PBDMA ..... Maleated low molecular weight polybutadienes
PE-g-MA .... Graft-copoly (ethylene/maleic anhydride)
PEG .... Polyethyleneglycol
PET .... Polyethylene terephthalate
PP .... Polypropylene
PP/ODPTA . Co-poly-(paraphenylene/3,4'-oxydiphenylene terephthalamide)
PPTA .... P-aramid poly(p-phenylene terephtalamide)
RFL .... Resorcinol formaldehyde latex fibre dip
SAXS .... Small angle X-ray diffraction
SBR .... Styrene-butadiene rubber
SBS .... Styrene-butadiene-styrene copolymer
SEM .... Scanning electron microscopy
SIS .... Styrene-isoprene-styrene copolymers
Stan .... Standard oily fibre finish
TBBS .... Tertiary-Butyl-BenzothiazolSulfenamide
TFA .... Tear fatigue analyser
TMQ 1,2-dihydro-2,2,4-trimethylquinoline
TPE .... Thermoplastic elastomer
TPNR .... Thermoplastic natural rubber
TPU .... Thermoplastic polyurethane
TRIM .... Trimethylol-propane-trimethacrylate
UR .... Urethane
XNBR .... Carboxylated nitrile rubber
XPS .... X-ray photoelectron spectroscopy

List of Figures

List of Tables

Bibliography

[1] D. H. Meadows, J. Randers, and D. L. Meadows. *Limits to growth: the 30-year update.* Earthscan, 2005.

[2] Wackernagel et al. Tracking the ecological overshoot of the human economy. *Proceedings of the National Academy of Sciences USA*, 99:9266–9271, 2002. 14.

[3] N.N. Wdk-branchenbericht. *Gummi Fasern Kunststoffe*, 6:372, 2012.

[4] finanzen.net. Brent oil price in us-dollar, April 2012. *http : //finanzen.net/rohstoffe/oelpreis/Chart.*

[5] European Commission Energy. Labelling of tyres, 2011. *http : //ec.europa.eu/energy/efficiency/labelling/labelling$_o$f$_t$yres$_e$n.htm.*

[6] S. Lange. Green tire materials. *Presentation from Continental AG on 4th Dresden Tire Workshop 6.5.2011*, 2011.

[7] J.F. Van der Pol. Short para aramid fiber reinforcement. *Rubber World*, 32:210, 1994.

[8] M. Koschmieder. *Processing and Properties of Elastomer Matrix Composites.* Ph.d. thesis, RWTH Aachen, Germany, 2000.

[9] S. K. De and J. R. White. *Short fibre-polymer composites.* Woodhead Publishing, 1996.

[10] S. Ranganathan and S. G. Advani. Chapter 2 fiber-fiber and fiber wall interactions during the flow of non-dilute suspensions. In T. D. Papathanasiou and D. C. Guell, editors, *Flow-induced alignment in composite materials*, pages 43–71. Woodhead, Cambridge, 1997.

[11] R. S. Rajeev. Chapter 12 fiber reinforced elastomers. In A. K. Bhowmick, editor, *Current Topics in Elastomers Research*, pages 351–390. CRC Press, 2008.

[12] L. A. Goettler and W.F. Cole. Chapter 9 short fiber-filled rubber composites. In A. K. Bhowmick and H. L. Stephens, editors, *Handbook of elastomers*, pages 241–264. Marcel Dekker, New York, 2001.

[13] S. Abrate. The mechanics of short-fiber-reinforced composites - a review. *Rubber Chemistry and Technology*, 59(3):384–404, 1986.

[14] A. R. Kelly and W.R. Tyson. Tensile properties of fibre-reinforced metals: Copper/tungsten and copper/molybdenum. *Journal of the Mechanics and Physics of Solids*, 13(6):329–350, 1965.

[15] B. W. Rosen and N. F. Dow. Mechanics of failure of fibrous composites. *Fracture*, 7:612, 1972.

[16] J. E. O'Connor. Short fibre reinforced elastomer composites. *Rubber Chemistry and Technology*, 50:945–958, 1977.

[17] J. C. Halpin and Tsai S.C. Effects of environmental factors on composite materials. *AFML TR*, 67, 1969.

[18] C. W. Bert and R. A. Kline. Composite material mechanics properties of planar-random fiber composites. *Polymer Composites*, 6(3):133, 1985.

[19] J. L. White. *Rubber processing: technology, materials, principles.* Hanser Publishers, 1995.

[20] G. Heinrich and M. Klüppel. Recent advances in the theory of filler networking in elastomers. In *Filled Elastomers Drug Delivery Systems*, volume 160 of *Advances in Polymer Science*, pages 1–44. Springer Berlin / Heidelberg, 2002.

[21] A. R. Payne. The dynamic properties of carbon black-loaded natural rubber vulcanizates. part i. *Journal of Applied Polymer Science*, 6(19):57–63, 1962.

[22] J. A. C. Harwood, L. Mullins, and A. R. Payne. Stress softening in natural rubber vulcanizates. part ii. stress softening effects in pure gum and filler loaded rubbers. *Journal of Applied Polymer Science*, 9:3011–3021, 1965.

[23] L. Ibarra. Dynamic properties of short fiber-epdm matrix composites as a function of strain amplitude. *Journal of Applied Polymer Science*, 54(11):1721–1730, 1994.

[24] R. S. Rajeev, A. K. Bhowmick, S. K. De, G. J. P. Kao, and S. Bandyopadhyay. New composites based on short melamine fiber reinforced epdm rubber. *Polymer Composites*, 23(4):574–591, 2002.

[25] M. A. Manchado and M. Arroyo. Short fibers as reinforcement of rubber compounds. *Polymer Composites*, 23(4):666–673, 2002.

[26] R. N. Datta and B. Pierik. Improving cut/chip/chunk resistance by using sulfron 3000. *Kautschuk Gummi Kunststoffe*, 6:328, 2007.

[27] R. N. Datta, N. Huntink, M. V. Van der Made, and B. Pierik. Improved hysteresis and adhesion to steel cord by using chemically activated aramid fiber. *Kautschuk Gummi Kunststoffe*, 61(11):580–583, 2008.

[28] V. M. Murthy, A. K. Bhowmick, and S. K. De. Scanning electron microscopy studies of failure surfaces of short glass fibre rubber composites. *Journal of Materials Science*, 17(3):709–716, 1982.

[29] F. Cataldo. Study on the reinforcing effect of milled carbon fibers in a natural rubber based composite. *Journal of Macromolecular Science Part B Physics*, 47:818–828, 2008.

[30] V. G. Geethamma, K. T. Mathew, R. Lakshminarayanan, and S. Thomas. Composite of short coir fibres and natural rubber: effect of chemical modification, loading and orientation of fibre. *Polymer*, 39(6-7):1483–1491, 1998.

[31] S. Varghese, B. Kuriakose, S. Thomas, and K. Joseph. Effect of adhesion on the equilibrium swelling of short sisal fiber reinforced natural rubber composites. *Rubber Chemistry and Technology*, 86:37–49, 1994.

[32] V. M. Murty, S. K. De, S. S. Bhagawan, R. Sivaramakrishnan, and S. K. Athithan. Viscoelastic properties of short-fiber-reinforced rubber composites and the role of adhesion. *Journal of Applied Polymer Science*, 28(11):3485–3495, 1983.

[33] P.K. Bipinbal and K. N. Kutty. A comparative study of short nylon fiber - natural rubber composites prepared from dry rubber and latex masterbatch. *Journal of Applied Polymer Science*, 109:1484, 2008.

[34] K. Agarwal, D. K. Setua, and G. N. Mathur. Short fibre and particulate-reinforced rubber composites. *Defence Science Journal*, 52(3):337–346, 2002.

[35] C. Rajesh, G. Unnikrishnan, E. Purushothaman, and S. Thomas. Cure characteristics and mechanical properties of short nylon fiber reinforced nitrile rubber composites. *Journal of Applied Polymer Science*, 92(2):1023–1030, 2003.

[36] S. K. Chakraborty, D. K. Setua, and S. K. De. Short jute fiber reinforced carboxylated nitrile rubber. *Rubber Chemistry and Technology*, 55(5):1286–1307, 1982.

[37] N. Wada, K. Fukunaga, and Uchiyama Y. Dispersion and orientation of short fiber reinforced rubber composites. *Kautschuk, Gummi, Kunststoffe*, 44:1141, 1991. 12.

[38] D. K. Setua and S. K. De. Short silk fiber reinforced nitrile rubber composites. *Journal of Materials Science*, 19(3):983–999, 1984.

[39] M.C.H. Lee. Chapter 8: Design and applications of short fibre reinforced rubbers. In S. K. De and J. R. White, editors, *Short fibre-polymer composites*, pages 192–209. Woodhead Publishing, 1996.

[40] I. Orlob and T. Reußmann. Hochleistungs-elastomere auf basis von aramidkurzfaserverstärktem silikonkautschuk. *Gummi Fasern Kunststoffe*, 2:94–99, 2011.

[41] R. P. Kumar, K. C. M. Nair, S. Thomas, S. C. Schit, and K. Ramamurthy. Morphology and melt rheological behaviour of short-sisal-fibre-reinforced sbr composites. *Composites Science and Technology*, 60(9):1737–1751, 2000.

[42] U. Wangenheim and R. Schuster. *Herstellung und Charakterisierung von Nanofasern zur Verstärkung von Mikrobauteilen durch Electrospinning*. 15. DKG-Fachtagung am 23. und 24. Mai 2011, Fulda, Germany. 2011.

[43] S. Praveen, P. K. Chattopadhyay, S. Jayendran, B. C. Chakraborty, and S. Chattopadhyay. Effect of nanoclay on the mechanical and damping properties of aramid short fibre filled styrene butadiene rubber composites. *Polymer International*, 59(2):187–197, 2010.

[44] R. N. Datta. Reduced hysteresis in truck tread compounds by using aramid short aramid fibers. *Rubber Chemistry and Technology*, 79:26–38, 2006.

[45] M. R. Kashani. Aramid-short-fiber reinforced rubber as a tire tread composite. *Journal of Applied Polymer Science*, 113(2):1355–1363, 2009.

[46] B. Rijpkema. The use of short fibers to reduce the rolling resistance of tires. *Kautschuk Gummi Kunststoffe*, 47:748–752, 1994.

[47] M. Gunnarsson and J. Nyman. *Short fibre reinforced Elastomers.* Master thesis, Lund University, 1998.

[48] I. Ahmad, P. Y. Wong, and I. Abdullah. Effects of fiber composition and graft-copoly(ethylene/maleic anhydride) on thermoplastic natural rubber composites reinforced by aramid fiber. *Polymer Composites*, 27(4):395–401, 2006.

[49] I. Ahmad, T. S. Chin, C. K. Cheong, A. Jalar, and I. Abdullah. Study of fiber surface treatment on reinforcement/matrix interaction in twaron fiber/enr composites. *American Journal of Applied Sciences (Special Issue)*, pages 14–20, 2005.

[50] M. Arroyo, R. Zitzumbo, and F. Avalos. Composites based on pp/epdm blends and aramid short fibres. morphology/behaviour relationship. *Polymer*, 41(16):6351–6359, 2000.

[51] S. K. N. Kutty and G. B. Nando. Short kevlar fiber thermoplastic polyurethane composite. *Journal of Applied Polymer Science*, 43(10):1913–1923, 1991.

[52] C. Vajrasthira, T. Amornsakchai, and S. Bualek-Limcharoen. Fiber-matrix interactions in aramid-short-fiber-reinforced thermoplastic polyurethane composites. *Journal of Applied Polymer Science*, 87(7):1059–1067, 2003.

[53] M. Akbarian, S. Hassanzadeh, and M. Moghri. Short twaron aramid fiber reinforced thermoplastic polyurethane. *Polymers for Advanced Technologies*, 19:1894–1900, 2008.

[54] K Naskar and Shibulal G. S. Exploring a novel multifunctional agent to improve the dispersion of short aramid fiber in polymer matrix. *exPRESS Polymer Letters*, 6:329–344, 2012.

[55] K Naskar and Shibulal G. S. Rfl coated aramid short fiber reinforced thermoplastic elastomer: Mechanical, rheological and morphological characteristics. *Journal of Polymer Resins*, 18:2295–2306, 2011.

[56] L. A. Goettler, R. I. Leib, and A. J. Lambright. Short fiber reinforced hose - new concept in production and performance. *Rubber Chemistry and Technology*, 52(4):838–863, 1979.

[57] A. Y. Coran, K. Boustany, and P. Hamed. The properties of oriented cellulose fiber-elastomer composites. *Rubber Chemistry and Technology*, 47:396–1751, 1974.

[58] R. P. Kumar and S. Thomas. Tear and processing behaviour of short sisal fibre reinforced styrene butadiene rubber composites. *Polymer International*, 38(2):173–182, 1995.

[59] S. Akhtar, P. P. De, and S. K. De. Short fiber-reinforced thermoplastic elastomers from blends of natural rubber and polyethylene. *Journal of Applied Polymer Science*, 32(5):5123–5146, 1986.

[60] H. Ismail, N. Rosnah, and H.D. Rozman. Curing characteristics and mechanical properties of short oil palm fibre reinforced rubber composites. *Polymer*, 38(16):4059–4064, 1997.

[61] A.P. Foldi. Reinforcement of rubber through short individual filaments. *Rubber Chemistry and Technology*, 49:379–389, 1976.

[62] P. J. Derringer. Short fiber-elastomer composites. *Journal of Elastomers and Plastics*, 3:230, 1971.

[63] W. Guo and M. Ashida. Dynamic viscoelasticities for short fiber-thermoplastic elastomer composites. *Journal of Applied Polymer Science*, 50(8):1435–1443, 1993.

[64] J. E. Stamhuis. Mechanical properties and morphology of polypropylene composites. iii. short glass fiber reinforced elastomer modified polypropylene. *Polymer Composites*, 9(4):280–284, 1988.

[65] P. B. Jana, S. K. De, S. Chaudhuri, and A. K. Pal. Electrical-conductivity of barium-ferrite-vulcanized polychloroprene filled with short carbon-fiber. *Rubber Chemistry and Technology*, 65(1):7–23, 1992.

[66] S. R. Moghue. Mechanical properties of short fiber elastomer composites. *Rubber Chemistry and Technology*, 49:1160–1166, 1976.

[67] E. G. Chatzi and J. L. Koenig. Morphology and structure of kevlar fibers: A review. *Polymer-Plastics Technology and Engineering*, 26(3):229–270, 1987.

[68] H. H. Yang. *Kevlar aramid fiber*. John Wiley and Sons, 1993.

[69] K. Tashiro, M. Kobayashi, and H. Tadokoro. Elastic moduli and molecular structures of several crystalline polymers, including aromatic polyamides. *Macromolecules*, 10(2):413, 1977.

[70] K. Haraguchi, T. Kajiyama, and M. Takayanagi. *Ibid*, 23:903, 1979.

[71] M. Panar, P. Avakian, R. C. Blume, K. H. Gardner, T. D. Gierke, and H. H. Yang. Morphology of poly(p-phenylene terephthalamide) fibers. *Journal of Polymer Science: Polymer Physics Edition*, 21(10):1955–1969, 1983.

[72] M. G. Dobb, D. J. Johnson, and B. P. Saville. *Journal of Polymer Science: Polymer Physics Edition*, page 2201, 1977.

[73] H. Aoki, D. R. Coffin, T. A. Hancock, D. Harwood, R. S. Lenk, John F. Fellers, and J. L. White. Synthesis, characterization, rheological, and fiber formation studies of p-linked aromatic polyamides. *Journal of Polymer Science: Polymer Symposia*, 65(1):29–40, 1978.

[74] N.N. Teijin aramid pulp and chopped fiber, 2004.

[75] L. Konopasek and J. W. S. Hearle. The tensile fatigue behavior of para-oriented aramid fibers and their fracture morphology. *Journal of Applied Polymer Science*, 21(10):2791–2815, 1977.

[76] M. E. G. Mosquera, M. Jamond, A. Martinez-Alonso, and J. M. D. Tascon. Thermal transformations of kevlar aramid fibers during pyrolysis: Infrared and thermal analysis studies. *Chemistry of Materials*, 6(11):1918–1924, 1994.

[77] M. G. Dobb, R. M. Robson, and A. H. Roberts. The ultraviolet sensitivity of kevlar 149 and technora fibres. *Journal of Materials Science*, 28(3):785–788, 1993.

[78] Dupont. Technical guide kevlar aramid fiber, 2000.

[79] H. Watanabe, M. Furukawa, T. Takata, and M. Yamamoto. Surface improvements of aramid fibers by physical treatments. *Macromolecular Symposia*, 159(1):131–142, 2000.

[80] J. W. Cherrie, H. Gibson, C. Mclntosh, W. M. Maclaren, and Lynch G. Exposure to fine airborne fibrous dust amongst processors of para aramid. *The Annals of Occupational Hygiene*, 39(4):403–125, 1995.

[81] J. P. Marsh, B. T. Mossman, K. E. Driscoll, R. F. Schins, and P. J. A. Borm. Effects of aramid, a high strength synthetic fiber, on respiratory cells in vitro. *Drug and Chemical Toxicology*, 17(2):75–92, 1994.

[82] M. C. Andrews, D. Lu, and R. J. Young. Compressive properties of aramid fibres. *Polymer*, 38(10):2379–2388, 1997.

[83] S. J. Deteresa, S. Allen, R. Farris, and R. Porter. Compressive and torsional behaviour of kevlar 49 fibre. *Journal of Materials Science*, 19(1):57–72, 1984.

[84] M. G. Dobb, D. J. Johnson, and B. P. Saville. Compressional behaviour of kevlar fibres. *Polymer*, 22(7):960–965, 1981.

[85] A. Hashimoto, M. Satoh, T. Iwasaki, and Morita M. Fibrillation of aramid fiber using a vibration ball mill and evaluation of the degree of fibrillation. *Journal of Materials Science*, 37:4013–4017, 2002.

[86] G. Roebroeks and W. H. M. Dreumel. Mechanical interfacing of aramid fibre. In K. Brunch, H. D. Gölden, and Herkert C. M, editors, *High Tech - the way into the nineties*, pages 95–102. Elsevier Science Publishers B.V, 1986.

[87] T. F. Schuler. Kevlar short fiber reinforcement of elastomer matrices. *Kautschuk Gummi Kunststoffe*, 7:548–549, 1992.

[88] P. J. de Lange, P. G. Akker, S. Willemsen, and R. N. Datta. The effect of oily finish components on the adhesion between aramid fibers and rubber. *Journal of Adhesion Science and Technology*, 23:139–149, 2009.

[89] P. J. de Lange, E. Maeder, K. Mai, R. J. Young, and I. Ahmad. Characterization and micromechanical testing of the interphase of aramid-reinforced epoxy composites. *Composites Part A: Applied Science and Manufacturing*, 32(3-4):331–342, 2001.

[90] P. J. de Lange, P. G. Akker, E. Maeder, S. Gao, W. Prasithphol, and R. J. Young. Controlled interfacial adhesion of twaron aramid fibres in composites by the finish formulation. *Composites Science and Technology*, 67(10):2027–2035, 2007.

[91] W. B. Wennekes. *Adhesion of RFL-treated cords to rubber: new insights into interfacial phenomena.* Ph.d. thesis, University of Twente, The Netherlands, 2008.

[92] W. B. Wennekes, J.W.M. Datta, R.N. Noordermeer, and F. Elkink. Fibre adhesion to rubber compounds. *Rubber Chemistry and Technology*, 60:523–540, 2008.

[93] Sartomer. Enhancement of the properties and dispersion of chopped aramid fiber loaded rubber compositions with application of maleinized low molecular weight polybutadienes, 2008.

[94] N. Inagaki, S. Tasaka, and H. Kawai. Surface modification of aromatic polyamide film by aminoethanethiol solution for silicon rubber composites. *Journal of Applied Polymer Science*, 56(6):677–686, 1995.

[95] A. B. Coffey, C. M. O'Bradaigh, and R. J. Young. Interfacial stress transfer in an aramid reinforced thermoplastic elastomer. *Journal of Materials Science*, 42(19):8053–8061, 2007.

[96] M. M. Hassan, G. A. Mahmoud, H. H. El-Nahas, and E. S. A. Hegazy. Reinforced material from reclaimed rubber/natural rubber, using electron beam and thermal treatment. *Journal of Applied Polymer Science*, 104(4):2569–2578, 2007.

[97] P. J. de Lange and P. G. Akker. Adhesion activation of twaron aramid fibers: Chemical versus plasma treatment. In *International Conference on Interfaces and Interphases in Multicomponent Materials*, The University of Sheffield, Sheffield, UK, 2010.

[98] F. Poncin-Epaillard, B. Chevet, and J. Brosse. Study of an aramid surface reactivity: Modification with a cold plasma or an electron beam followed by a postgrafting reaction. *Journal of Applied Polymer Science*, 52(8):1047–1061, 1994.

[99] S. C. Sharma. Adhesion of aramid cords to rubber. *US4469748*, 1984.

[100] K. W. Stöckelhuber, A. Das, R. Jurk, and G. Heinrich. Contribution of physico-chemical properties of interfaces on dispersibility, adhesion and flocculation of filler particles in rubber. *Polymer*, 51(9):1954–1963, 2010.

[101] R. Pristavok. *Analyse und Modellierung der Haftungsmechanismen bei der Beschichtung und Verklebung von Papierwerkstoffen.* Ph.d. thesis, Technischen Universität Bergakademie Freiberg, Germany, 2006.

[102] Young T. Cohesion of fluids. *Phil. Trans.*, 95:65, 1805.

[103] A. Dupré. *Theorie mecanique de la chaleur.* Paris 369, 1869.

[104] F.M. Fowkes. *Journal of Physical Chemistry*, 67:2538, 1963.

[105] J. L. White and K. Min. Towards understanding of the rheology of rubber compounds and their processing. *Rubber Chemistry and Technology*, 82:131–148, 2009.

[106] N.N. *DIN 53523 Prüfung mit Scherscheiben-Viskosimeter nach Mooney*. Prüfung von Kautschuk und Elastomeren. Berlin, 1991.

[107] L. A. Goettler and K. S. Shen. Short fibre reinforced elastomers. *Rubber Chemistry and Technology*, 56:619–638, 1983.

[108] M. Ashida. Chapter 5: Composites of polychloroprene rubber with short fibres of poly(ethylene terephthalate) and nylon. In S. K. De and J. R. White, editors, *Short fibre-polymer composites*, pages 116–143. Woodhead Publishing, 1996.

[109] G. Ortlepp. Bündelfreie kurzfasern für neue anwendungen in thermoplasten und elastomeren. *Gummi Fasern Kunststoffe*, 4:246–248, 2008.

[110] P. B. Jana, S. Chaudhuri, A. K. Pal, and S. K. De. Electrical-conductivity of short carbon fiber-reinforced polychloroprene rubber and mechanism of conduction. *Polymer Engineering and Science*, 32(6):448–456, 1992.

[111] K. S. Shen and R. K. Rains. Scale-up of short fiber composite mixing. *Rubber Chemistry and Technology*, 52(4):764–772, 1979.

[112] M. M. H. Kuroda and C. E. Scott. Blade geometry effects on initial dispersion of chopped glass fibers. *Polymer Composites*, 23(5):828–838, 2002.

[113] M. M. H. Kuroda and C. E. Scott. Initial dispersion mechanisms of chopped glass fibers in polystyrene. *Polymer Composites*, 23(3):395–405, 2002.

[114] J. L. White and K. Min. Flow visulaization investigations of the addition of carbon black and oil to elastomers in an internal mixer. *Rubber Chemistry and Technology*, 60:361–380, 1980.

[115] L. A. Goettler, A. J. Lambright, R. I. Leib, and P. J. Dimauro. Extrusion-shaping of curved hose reinforced with short cellulose fibers. *Rubber Chemistry and Technology*, 54(2):277–301, 1981.

[116] J. Clarke and J. Harris. Controlled orientation of short fibre reinforcement for anisotropic performance of rubber compounds. *Plastics, Rubber and Composites*, 30:406–415, 2001.

[117] J.L. Thomason and M.A. Vlug. Influence of fibre length and concentration on the properties of glass fibre-reinforced polypropylene: 1. tensile and flexural modulus. *Composites Part A: Applied Science and Manufacturing*, 27(6):477–484, 1996.

[118] C.Y. Park and B.K. Kim. *Pollimo*, 14:197, 1990.

[119] A.P. Foldi. Chapter 9: Design considerations and end-use applications of short fibre filled rubbers and thermoplastic elastomers. In S. K. De and J. R. White, editors, *Short fibre-polymer composites*, pages 210–256. Woodhead Publishing, 1996.

[120] R. P. Kumar and S. Thomas. Short fibre elastomer composites: effect of fibre length, orientation, loading and bonding agent. *Bulletin of Materials Science*, 18(8):1021–1029, 1995.

[121] M. Ashida, T. Noguchi, and S. Mashimo. Dynamic moduli for short fiber-cr composites. *Journal of Applied Polymer Science*, 29(2):661–670, 1984.

[122] M. Ashida, T. Noguchi, and S. Mashimo. Effect of matrix's type on the dynamic properties for short fiber-elastomer composite. *Journal of Applied Polymer Science*, 30(3):1011–1021, 1985.

[123] Y. Uchiyama, N. Wada, T. Iwai, S. Ueda, and S. Sado. Friction of short-fiber-reinforced rubber on wet surfaces. *Journal of Applied Polymer Science*, 95(1):82–89, 2005.

[124] N. Wada, Y. Uchiyama, and M. Hosokawa. Friction of fibrious fillers and their orientations on the wear of short fibre reinforced rubber composites. *International Polymer Science and technology*, 20:8, 1993.

[125] N. Wada and Y. Uchiyama. Effect of surface roughness and fibre diameter on the abrasive wear of short fibre reinforced rubber composites. *International Polymer Science and technology*, 21:1, 1994.

[126] R. J. Young, D. J. Bannister, A. J. Cervenka, and I. Ahmad. Effect of surface treatment upon the pull-out behaviour of aramid fibres from epoxy resins. *Journal of Materials Science*, 35(8):1939–1947, 2000.

[127] H. B. Shen, S. Nutt, and D. Hull. Direct observation and measurement of fiber architecture in short fiber-polymer composite foam through micro-ct imaging. *Composites Science and Technology*, 64(13–14):2113–2120, 2004.

[128] F. Sieker and O. Brunke. Computed tomography for analysis of rubber components. *Kautschuk Gummi Kunststoffe*, 5:224–226, 2009.

[129] T. Alshuth and S. Robin. Hochauflösende 3d röntgen computertomographie. *Kautschuk Gummi Kunststoffe*, 9:383–387, 2010.

[130] D. Shaw. Trelleborg uses aramid pulp in hoses. *European Rubber Journal*, 190(6):22, 2008.

[131] R. J. Brown and Scriver R. M. Rubber containing aramid pulp reinforcement. *US4871004*, 14, 1989.

[132] B. Nanni, M. N. Antonio and C. Zanichelli. Tire, including fibre reinforced elastomeric intermediate layer between the belt structure and tread band, and method of making the tire, 2004.

[133] C. H. Causa, A.G. Obermaier and M. Borowczak. Elastomer containing partially orientated fibres, tires made using said elastomers and a method therefor. *EP0691218A2*, 30, 1996.

[134] B. A. Bergman and Arbold J. J. Elastomer composition having glass micro fibers. *PCT/US2006/034065*, 25, 2008.

[135] E. Görter, G. Kresta and H. Stumpf. Molded article formed of fibre reinforced material, 1974.

[136] K. Boustany and A. Y. Coran. A pneumatic tire containing high modulus discontinous fibre reinforced member. *US3802478*, 1972.

[137] K. Boustany and A. Y. Coran. Discontinuous cellulose reinforced elastomer. *US3697364*, 17, 1972.

[138] D. Moers and M. Balland. Pneumatic tire with anisotropic filler strip. *US4236563*, 4, 1980.

[139] G. Guichard and A. Pouchelon. Reinforced silicone elastomer. *PCT/EP2007/064122*, 30, 2008.

[140] N.N. Keltan 8340a datasheet, 2004.

[141] S. Fu, Y. Mai, E. C. Ching, and R. K. Y. Li. Correction of the measurement of fiber length of short fiber reinforced thermoplastics. *Composites Part A: Applied Science and Manufacturing*, 33(11):1549–1555, 2002.

[142] N.C Davidson and A.R. Clarke. Extending the dynamic range of fibre length and fibre aspect ratios by automated image analysis. *Journal of Microscopy*, 196(2):266–272, 1999.

[143] N.P. Cheremisinoff. *Polymer Mixing and Extrusion Technology*. Plastics Engineering. Taylor & Francis, 1987.

[144] A. I. Medalia. Microscopic estimation of carbon black dispersion. *Rubber Age*, 97:82, 1965.

[145] A. R. Clarke, G. Archenhold, N. C. Davidson, W. S. Slaughter, and N. A. Fleck. Determining the power spectral density of the waviness of unidirectional glass fibres in polymer composites. *Applied Composite Materials*, 2(4):233–243, 1995.

[146] S. Pegel, P. Pötschke, T. Villmow, D. Stoyan, and G. Heinrich. Spatial statistics of carbon nanotube polymer composites. *Polymer*, 50(9):2123–2132, 2009.

[147] N.N. *DIN 53529 Vulkametrie*. Prüfung von Kautschuk und Elastomeren. Berlin, 1983.

[148] S. Wu. *Polymer interface and adhesion*. Marcel Dekker New York, 1982.

[149] Kaelble D.H. Dispersion-polar surface tension properties of organic solids. *Journal of Adhesion*, 2:66, 1970.

[150] D.K. Owens and Wendt R.C. Estimation of the surface free energy of polymers. *Journal of Applied Polymer Science*, 13:1741–1747, 1969.

[151] N.N. *DIN 53504 Bestimmung von Reißfestigkeit, Zugfestigkeit, Reißdehnung und Spannungswerten im Zugversuch*. Prüfung von Kautschuk und Elastomeren. Berlin, 2009.

[152] N.N. *DIN 53505 Härteprüfung nach Shore A und Shore D*. Prüfung von Kautschuk und Elastomeren. Berlin, 2000.

[153] M. L. Williams, R. F. Landel, and J. D. Ferry. The temperature dependence of relaxation mechanisms in amorphous polymers and other glass-forming liquids. *Journal of American Chemistry Society*, 77:3701–3707, 1955.

[154] N.N. *DIN 53533 Prüfung der Wärmebildung und des Zermürbungswiderstandes im Dauerschwingversuch (Flexometerprüfung)*. Prüfung von Kautschuk und Elastomeren. Berlin, 1988.

[155] R. Stocek. *Dynamischen Rissausbreitung in Elastomerwerkstoffen*. Ph.d. thesis, Technische Universität Chemnitz, Germany, 2012.

[156] R. Stocek and G. Heinrich. Analyse des dynamischen risswachstums von elastomeren. *Kautschuk Gummi Kunststoffe*, 4:170–176, 2009.

[157] R. S. Rivlin and A. G. Thomas. Rupture of rubber. i. characteristic energy for tearing. *Journal of Polymer Science*, 10(3):291–318, 1953.

[158] P. C. Paris and F. Erdogan. A critical analysis of crack propagation laws. *Journal of Basic Engineering, Transactions of the American Society of Mechanical Engineers*, pages 528–534, 1963.

[159] N.N. Communication with teijin aramid, 2011.

[160] N.N. Keltan 2340a datasheet, 2004.

[161] D. J. Angier and W. F. Watson. Mastication of rubber. ii. interpolymerization of natural rubber and neoprene on cold milling. *Journal of Polymer Science*, 18(87):129–140, 1955.

[162] N.N. Communication with dsm elastomers, 2011.

[163] E. Ürögiova, G. Heinrich, N. Probst, and Grivei E. Elastomers filled with metal coated carbon black. *Kautschuk Gummi Kunststoffe*, 9:455, 2008.

[164] K. Subramaniam, A. Das, and G. Heinrich. Development of conducting polychloroprene rubber using imidazolium based ionic liquid modified multi-walled carbon nanotubes. *Composites Science and Technology*, 71(11):1441–1449, 2011.

[165] G. Heinrich. Dynamics of carbon black filled networks, viscoelasticity and wet skid behavior. *Kautschuk Gummi Kunststoffe*, 45(3):173–179, 1992.

[166] S. J. Deteresa, R. Farris, and R. Porter. Experimental verification of a microbuckling model for the axial compressive failure of high performance polymer fibres. *Journal of Materials Science*, 23(5):1886–1894, 1988.